MAKi SENSE

A STUDENT'S GUIDE TO RESEARCH AND WRITING

PSYCHOLOGY AND THE LIFE SCIENCES
FOURTH EDITION

MARGOT NORTHEY / BRIAN TIMNEY

OXFORD
UNIVERSITY PRESS

OXFORD
UNIVERSITY PRESS

70 Wynford Drive, Don Mills, Ontario M3C 1J9
www.oup.com/ca

Oxford University Press is a department of the University of Oxford.
It furthers the University's objective of excellence in research, scholarship,
and education by publishing worldwide in

Oxford New York
Auckland Cape Town Dar es Salaam Hong Kong Karachi
Kuala Lumpur Madrid Melbourne Mexico City Nairobi
New Delhi Shanghai Taipei Toronto

With offices in
Argentina Austria Brazil Chile Czech Republic France Greece
Guatemala Hungary Italy Japan Poland Portugal Singapore
South Korea Switzerland Thailand Turkey Ukraine Vietnam

Oxford is a trade mark of Oxford University Press
in the UK and in certain other countries

Published in Canada
by Oxford University Press

Copyright © Oxford University Press Canada 2007

The moral rights of the author have been asserted

Database right Oxford University Press (maker)

First published 2007

Library and Archives Canada Cataloguing in Publication

Northey, Margot, 1940–
Making sense : a student's guide to research and writing : psychology and
the life sciences / Margot Northey, Brian Timney. — 4th ed.

First-2nd eds. published under title: Making sense in psychology and the life sciences.
Includes bibliographical references and index.
ISBN-13: 978-0-19-542621-2
ISBN-10-19-542621-5

1. Psychology—Authorship. 2. Life sciences—Authorship. 3. Report writing.
I. Timney, Brian II. Northey, Margot, 1940– Making sense in psychology and the
life sciences. III. Title.

BF76.7.N67 2007   808'.06615   C2006-906258-7

2 3 4 – 10 09 08 07
This book is printed on permanent (acid-free) paper ∞.
Printed in Canada

TABLE OF CONTENTS

PREFACE

Although the rules of writing stay much the same, a great deal has changed with respect to our ability to gain access to research materials and to prepare documents using computers. When the first edition of this book was published in 1986, the IBM personal computer was less than five years old and the Apple Macintosh less than two. There were no laptops. The Internet did not exist, and almost every research resource available in a library was a hard copy. Only a small proportion of students had everyday access to a personal computer, so most papers were either typed or handwritten. By the time the second edition came out in 1994, most students were familiar with computers, but the prime use for PCs was word processing. Now, students use computers routinely as an integral part of their academic program.

This fourth edition of *Making Sense in Psychology and the Life Sciences* is written in the context of this computer-oriented world. New material on efficiently using the Internet as a research tool, and on evaluating online resources for accuracy, reliability, and timeliness has been included. Other chapters have been revised to take into account the fact that computers can be used for many other tasks than just word processing. Chapter 9, on the presentation of data, now provides more detailed examples of the ways in which data can be presented. We have also incorporated the new requirements of the fifth edition of the *Publication Manual of the American Psychological Association* and the recently published seventh edition of *The CSE Manual for Authors, Editors, and Publishers*, including guidelines for the citation of online documents and websites. Because of the importance of organizing research and lecture notes, and of maintaining accurate data records, we have devoted a complete chapter to this topic. The ability to present a seminar or another type of formal presentation is extremely useful, both at university or college and out in the world of business. For this reason, we have included a chapter with guidelines on how to give effective presentations, including the use of graphic presentation software, such as PowerPoint. We have also included chapters on writing a book report, a research proposal, or thesis, and writing a resume and cover letter.

I'd like to thank Kathleen McGill, of Oxford University Press, who invited me to prepare this new edition. The request was a timely one and gave me the opportunity to discuss the technical innovations that have changed the way that students approach their academic work. In preparing the revisions, I have been guided by the comments of students and colleagues who have used this book previously. I'm very grateful for their suggestions, and I hope that I've addressed some, if not all, of the omissions from the previous books. I would also like to thank Jessica Coffey and Eric Sinkins for their editorial work on the manuscript. Although I prepared this book to help others write well, I am grateful to know that there others who can help me to improve my own writing.

Finally, I thank my daughter, Meagan, whose direct experience as a university student gave me added insight into what I needed to include in this book. She was also able to give me first-hand advice on the use of Web browsers and "surfing the Net" for information. This book is dedicated to her.

Brian Timney
London, Ontario

Symbols for common errors

NOTE: If any of the following markings appear on one of your essays or reports, consult Chapter 16 or 17 or Glossary II for help.

agr	agreement of subject and verb
amb	ambiguity
awk	awkwardness
cap	capitalization
cs	comma splice
dang	dangling modifier (*or* dm)
D	diction
gr	grammar or usage
mod	misplaced modifier
¶	new paragraph
//	parallelism
ref	pronoun reference
P	punctuation
quot	quotation marks
rep	repetition
RO	run-on sentence
frag	sentence fragment
ss	sentence structure
sp	spelling
sp inf	split infinitive
sub	subordination
T	tense
trans.	transition
⌣	transpose (change order of letters or words)
wdy	wordy
ww	wrong word

chapter 1

COMMUNICATING IN THE SCIENCES

"Anamander believed that we were desended from fosels. He believed that our aunt sisters were fish. He believed we were animals, aquatics, and had husks and hunted on land and under fins. He also believed that we came on land and shaged [sic] off, lost are husks and hard shells and got legs. . . ."

—excerpt from a student's examination answer on the kinds of evidence Darwin drew upon when formulating his theory of evolution

Under the pressure of writing an exam, we do not always produce our best work, but as the above quotation shows, there are limits to what is acceptable. Anaximander—not "Anamander"—was a pre-Socratic Greek philosopher who probably knew nothing about fossils and certainly did not suggest that we "lost are husks and hard shells and got legs." Even if he did, his beliefs have nothing to do with the *evidence* that Darwin used. Apart from its general incoherence, the reason this answer received a low grade was that the student did not address the question asked. This is one of the most serious errors students make when trying to answer questions set by their instructors.

Over the years, science has developed a set of implicit and explicit rules for scientific writing. Minor details may vary from one discipline to another, but it should be possible for a physicist or a psychologist to read a biology paper and judge whether the science is reasonable, even if he or she doesn't understand all the technical details. The challenges of presenting research findings are not the same as they are when you have to write, say, an English paper on how Edgar Allan Poe's various substance addictions contributed to his poetry. While there is room for creativity in scientific writing, there are also constraints. For instance, when you are writing a lab report, you will be required to follow certain guidelines. These guidelines may be provided by your professor or lab instructor, or they may be ones that are traditional for a particular discipline.

Having a clearly defined format makes your writing task easier. However, you are still responsible for all of the other aspects of the writing process. Start with a decision as to what you are going to write about and how you are going to organize your thoughts before you even begin to write. One of the most frustrating experiences you may experience as a student occurs when you get back an essay or lab report and find that the grade is lower than you had expected. When you ask your instructor why you received such a low grade, he or she might say something like, "It didn't hang together very well," or "You seemed to know the material, but you didn't get your points across as well as some other students." Often an instructor will not go into detail about how you might have improved your paper, so you are thrown back onto your own devices. This scenario can be especially irritating for the science student, who may have less opportunity to write papers than a student in arts and humanities or the social sciences. It may seem unfair that someone who knows less about the material than you do gets a better mark because the instructor seems to give unnecessarily strong emphasis to "writing style."

In fact, many instructors do place a premium on style, so it is important to realize that there is more to good writing than just correct grammar and spelling. These things make a difference, but it is also important that you express your ideas in a clear and logical way. Your instructors are not mind-readers; you cannot leave out essential parts of an argument and still expect to get credit. Nor should you take it for granted that just because your instructors know the topic, they will understand the ideas you are trying to communicate. They may, but their job is to evaluate what you have actually written, not what they think you meant to say.

In scientific writing it is especially important to express yourself clearly because much of the time you will be trying to convey complex information. To convince your reader that you know what you're talking about, you have to ensure that he or she can understand what you've written. The best science writers have the ability to describe a complicated theory or procedure in a way that not only makes it easy to understand but also conveys their enthusiasm for the material. This is not a skill that some people are just born with. Although not everyone can write like Stephen Jay Gould or Oliver Sacks, almost anyone can learn to write in a way that will convince the reader that he or she is scientifically literate and has a good grasp of the subject.

The main purpose of this book is to give you the tools to develop your scientific writing skills. Because writing in science typically requires you to deliver information in a particular way, you should always begin by considering the needs of the scientific reader. If you know what your reader will be looking for you are likely to produce a better paper. In what follows we will provide you with suggestions that will make your papers suitable for a scientific audience. We will introduce you to some of the conventions of scientific communication and show you how to follow them in order to produce an interesting and effective piece of work. We will also provide you with suggestions to help you improve any kind of writing or presentation you will have to do.

SCIENTIFIC THINKING AND SCIENTIFIC WRITING

Whether you are writing an essay for an English course, a review paper for a microbiology course, or a lab report for a psychology course, you are not likely to produce clear writing unless you have first done some clear thinking. This means first working out what you want to say, then organizing your thoughts before you commit them to paper.

For the most part, the ability to write clearly is independent of subject material and form, but there are occasional exceptions. Sometimes we encounter a student who produces excellent essays but disastrous lab reports. Different rules apply to different kinds of writing, and it is important to be aware of these rules so that your essays will be as strong as your lab reports, research papers, and presentations.

In some ways, however, the distinction between the kinds of writing required to complete an essay for a history course and a lab report for a biology course is artificial. After all, each is designed to tell a story based on evidence that has been gathered, sorted, and evaluated in a logical, systematic way. The only differences lie in the kind of evidence that is examined and the specific principles used to evaluate it. In fact, it is possible to be scientific in your approach to almost any academic discipline; all you have to do is ensure that your evidence is complete and your analysis of it is systematic and logical.

On the other hand, it is also possible to approach even "scientific" subject matter in a way that is anything but scientific. Perhaps the best way to see the difference between scientific and unscientific writing is to compare the descriptions of a particular finding in a scholarly journal and a tabloid newspaper. In a tabloid, claims by scientists are typically presented as established facts, with no attempt to evaluate how the data were gathered or whether the conclusions are justified. The story may be written in correct English, but no science instructor would be satisfied with it.

For example, a newspaper story on a new drug treatment for cancer might suggest that this is the cure everyone has been searching for. If you were to look up the original report in a medical journal, however, you might find that the treatment worked only for certain types of cancer, that the number of patients whose condition improved was quite small, and that the authors put all kinds of qualifiers on their conclusions. Although the newspaper article might lead you to think that a cure for cancer had been found, the journal article would likely lead you to conclude that this was just another small step in ongoing cancer research. What this example shows is that to be scientific, a report must not only present the author's own evaluation of the evidence but also provide enough information for readers to draw their own conclusions.

WHERE DO YOU START?

Even if you have been assigned a topic for your research report or essay, you may have difficulty just getting started. We've all stared at a blank page for much more time than we would have liked to, thinking: *What am I going to write?* In some cases, that initial paralysis we think of as "writer's block" turns into an excuse to procrastinate until there is only enough time left to just throw together a paper with minimal planning.

In cases such as this, it may be worthwhile to step back a little before you put your fingers to the keyboard. If you've done some background reading, you should already have an idea of what information the paper will contain; your problem is deciding what material to include and in what order. Some people find that if nothing springs to mind immediately, it can help to take some leisure time to think about the assignment without any pressure. Going for a walk or a run, going to the gym, or even just sitting quietly away from your desk will give you a chance to let ideas percolate and coalesce. Relaxation in this context does not mean watching TV or going out with friends; you need some quiet time to reflect on what your paper is going to look like. You may not always have a major inspiration, but most of the time you'll find a starting point.

INITIAL STRATEGIES

Before you even consider what you will write, you should ask yourself a number of questions. The answers to these questions will determine not only *what* you write but *how* you will write it. If you do not ask these questions before you start writing, you run the risk of producing a paper that lacks a clear focus and direction, and this will result in a lower grade.

WHAT IS THE PURPOSE OF THIS PIECE OF WRITING?

Two related issues that you should consider before you begin to write are the purpose of your writing and the approach you should take to achieve this purpose. Writing a review of a body of literature requires a different approach from the one you would take to write a paper considering the merits of two opposing theories. Sometimes you will be asked to *discuss* a particular topic, sometimes to *compare* different theoretical viewpoints. At other times you may be writing a lab report or a thesis. Each of these will require a different approach, but before you can settle on an approach, you must have a clear idea of the purpose of your assignment.

Even if you select your own topic, you must decide what you are trying to accomplish. Depending on the assignment, your purpose may be any one or more of the following:

- to describe and interpret an experiment you have done;
- to show that you can do independent library research;
- to demonstrate your ability to evaluate primary or secondary sources;
- to show that you understand certain terms, concepts, or theories;
- to demonstrate your knowledge of a topic;
- to show that you can think clearly and critically.

Although there is some overlap among these purposes, an assignment designed to see if you have read and understood specific material will certainly call for a different approach from one that is meant to test your critical thinking or research skills. Starting to work on an assignment without setting a goal is like setting out on a journey with no destination: your writing will be aimless, and you will never know when you have arrived at the end.

WHAT APPROACH SHOULD I TAKE?

The way you approach your paper will be determined in part by your purpose and also by the context of the course you are taking. How did your instructor describe the purpose of the course at the beginning of classes? What aspects of the course materials have been emphasized by your instructor, and how has he or she approached them? How does this particular assignment relate to key concepts and themes of the course? Has your instructor been mainly descriptive, or analytical, or critical? Your instructor's approach is probably a reflection of his or her own preferences in regard to the subject matter.

WHO IS MY *REAL* AUDIENCE?

Even if the only person who will read your paper is your instructor, don't think of that one individual as your target audience. If you do, you're likely to leave out important details of an explanation because you assume your instructor will know them already and will understand your explanation without this essential information. Instead of writing with your instructor in mind, try to think of your reader as someone who is knowledgeable in the discipline but doesn't know everything about your specific topic. The person who reads your paper can read only what is on the page, not what is in your head. Don't take detailed knowledge about your topic for granted. Remember, your task is to convince the reader that you know what you are talking about and that your arguments have merit.

Thinking about the reader also means taking into account the intellectual context in which he or she operates. If you were to write a paper on human sexuality for a biology course, it would be quite different from one on the same topic submitted for a psychology course. You have to make specific decisions about the background information you will supply, the terms you will need to explain, and the amount of detail that is appropriate for a given situation. When you are writing a lab report, you need to give much more procedural detail than you would if you were writing a review of the same topic. If you don't know who will be reading your paper—your professor, your tutorial leader, or a marker—just imagine someone intelligent, well informed, and interested, skeptical enough to question your ideas but flexible enough to accept them if your evidence is convincing.

If you are giving a seminar or an oral presentation, then you will have a real audience in front of you. You will have to take into account a much greater variation in background knowledge, so you have to be especially careful to be as clear as possible and to avoid leaving important details out of your descriptions and explanations.

HOW LONG SHOULD THE PAPER BE?

Before you start writing, you will also need to think about the length of the assignment in relation to the time available to you. If both the topic and the length are prescribed, it should be fairly easy for you to assess the level of detail required and the amount of research you need to do. If only the length is prescribed, that restriction will help you decide how broad or how narrow a topic you should choose. You should also keep in mind how much the assignment is worth. A paper that is worth 50 per cent of your final grade will need more of your time and effort than one that is worth only 10 per cent.

WHAT SHOULD THE TONE OF THE PAPER BE?

In everyday writing to friends you almost certainly take a casual tone, but academic writing is usually more formal. The exact degree of formality will depend on the kind of assignment and instructions you have been given. In some cases—for example, if your psychology professor asks you to keep a journal describing certain personal experiences—you may well be able to use an informal style.

However, in lab reports and review papers, where you need to express yourself unambiguously, a more formal tone is required.

On the other hand, you should also avoid the other extreme of excessive formality. You must resist the temptation to fill your work with long words and high-flown phrases, which will only make your writing sound stiff and pretentious. Finding a suitable tone for academic writing is a challenge for many students. In Chapter 14 we shall give you some guidelines for setting the tone of your paper.

COMPUTER LITERACY

It's a safe assumption today that every student has access to both a computer and to a wide variety of software for word processing, data analysis, graphics, and the preparation of visual presentations. Similarly, everyone now has access to the Internet. It is difficult to overestimate the role played by computers and the Internet in all areas of academic research. In the following chapters we will discuss how you can take advantage of this technology, both in the initial stages of your research for an assignment and in the final preparation of a paper or a talk.

A certain level of computer literacy, no matter what discipline you are studying, is essential. At the very least students will need to be familiar with a word processing program such as Word or WordPerfect. Familiarity with the Internet is becoming more and more essential. Many instructors use PowerPoint presentations to illustrate lectures, and they make the slides available to students on a course website. Increasingly, textbook publishers are creating supplementary websites for student use that might include practice tests, project outlines, and suggestions for further reading. Furthermore, the Internet—as we will discuss later—is often the first place to which many students turn when beginning their research.

Using a computer simplifies many of the tasks of writing: you can correct mistakes before they arrive on paper, experiment with the structure of your writing by cutting and pasting blocks of text, check your spelling, create a variety of tables of graphs, and prepare a clean and professional final copy. However, using a computer also presents a number of potential problems. Here are a few things to always keep in mind:

- **Don't let the system rule your thinking**. Seeing something typed out neatly on screen or on paper certainly looks professional. But don't be fooled into thinking that fancy graphics and a slick presentation can replace intelligent thinking. Thoughtful arguments, careful analysis, and clear organization will impress your professor more than the range of fonts and graphics used. Also, remember to carefully proofread your work—even when the program does not point out errors. While your program's spell-checker might catch obvious mistakes, it will not know that you meant to type "kind" instead of "kid" or "lead" instead of "led".

- **Save regularly and back up your files.** There is nothing worse than spending all day working on a paper only to lose everything because of a system crash or a power outage. While your computer may have an auto save function, it's a more reliable plan to simply take the time to save your work in regular intervals. It's also a good idea to occasionally save your work to a location other than your hard drive. This could mean saving it to a local network, a CD, or a memory stick. If you have regularly backed up your files in an external location, damage to your hard drive will have a minimal impact on your writing.

- **Don't discard your files.** You should be sure to keep a copy of your writing, at least until your paper has been graded and returned to you. The resolution of many cases of alleged plagiarism has been dependent on whether the student could produce an electronic version of the paper. Some universities now use specialized software to check for plagiarism, and may therefore require you to submit an electronic version of the paper as well as a printed copy. In this case, you should always keep an electronic version in case something happens to your instructor's copy.

chapter 2

DOING RESEARCH

Doing research means different things depending on your academic discipline. In this book we focus on two different kinds of research: the kind that you do in the library or on your computer, and the kind that you might do in a lab. Although they may seem very different, they have a great deal in common. If you think about the purpose of research you can see how this is the case. The goal of any research project is to answer a question that you have set yourself, and in order to do so, you will need to take a particular approach to the topic you are working on. It makes little difference whether you begin by reading and making notes from a book in the library, or by recording a person's responses to stimuli presented on a computer screen, or both; the same general rules will apply.

No matter what kind of research project you are doing you will follow the same three stages: obtaining information, organizing that information, and presenting the information to an audience. If you are writing an essay, you will obtain your information from the library or the Internet. If you are in a lab course, you may be conducting an experiment as well as doing some background reading. Although these are quite different activities, one thing that they share is the need for organization.

If you do a lot of reading, you will need to take notes in such a manner that they can be arranged in a variety of ways, depending on how you decide to organize your paper. If you do an experiment in the lab, you want to be sure that you have tabulated your data in a way that makes it easy to organize your analysis and the final write-up of your lab report. The goal of your research may be an essay, a lab report, or a thesis; it could also be a seminar or a more formal presentation at a conference. No matter what format you're using to present the results of your research, the rules are the same: you must be clear, concise, and coherent. In the rest of this chapter we will provide you with some guidelines that might help you achieve those goals.

ASKING QUESTIONS

The ability to ask intelligent questions is one of the most important—albeit underrated—skills that you can have, whether at university or in the outside world. Good academic research is all about asking the right questions. Sometimes, the most elegant research comes from asking some very simple questions. In the 1950s, Kenneth Roeder (1998), a neurobiologist, watched the ways that moths tried to avoid the bats that were hunting them. His hypothesis was that moths could hear and react to high-frequency sounds that bats emitted as they flew about. Beginning with a very simple experiment in which he recorded the responses of nerve cells in a moth's ear as a bat flew by, he went on to ask if there was a relationship between the proximity of the bat and the response. Then he asked if the nerve cells responded in a way that could identify the direction the bat was coming from. Each answer led him to another question, so that over a period of many years he was able to describe the whole predator-prey relationship between bats and moths.

What this example shows is that asking the right questions is an essential part of scientific investigation. Even if you are given specific instructions by your professor, you will still have to limit your project by posing a series of questions about your topic and how you should approach it. The kinds of questions you ask will depend in large part on the specific content of your work and on the approach you decide to take.

DEVELOPING ANSWERS

Research begins with asking questions. This is followed first by the collection of information, and then by the evaluation of that information in order to draw a conclusion. We will discuss methods of gathering information in the next section, but before that we will look at ways to draw conclusions from the information you have gathered.

One way to draw conclusions from the results of your research is by adopting the *scientific method*. You can think of the scientific method as a systematic way of approaching the questions you ask. In doing this you will use two of the basic principles of logical reasoning, *induction* and *deduction*.

Induction allows you to move from the study of individual cases to arrive at a general conclusion. In other words, you look for converging evidence that will support a hypothesis or theory. You should be aware, though, when taking this approach, that your conclusions will only be as reliable as the information they're based on. The reasoning process itself does not include any way to evaluate the representativeness of the specific cases you consider and has no built-in checks or controls.

Deduction begins with a general *a priori* premise—your hypothesis or theory—and moves through a process of evaluation or experimentation to allow you to explain particular events. Deductive reasoning is not the source of new information; rather, it is a way to uncover relationships as the researcher moves from the general to the specific. It depends on the assumption that the initial generalization is correct, which may not always be the case.

By themselves, induction and deduction do not establish truth or reliable knowledge. They are tools that allow you to evaluate the evidence available to you, and they operate best when used together in the context of the scientific method. The scientific method is a systematic approach to inquiry that relies on a set of orderly procedures, combining aspects of both induction and deduction to allow you to answer the questions you ask. Although the precise path you take will vary with the specific project you are working on, you would normally proceed in a logical fashion through a sequence of steps such as the following:

1. Formulating and delimiting the problem.
2. Reviewing the related literature.
3. Developing a theoretical framework.
4. Formulating hypotheses.
5. Selecting a research design.
6. Specifying the population that you will study.
7. Developing a plan for data collection.
8. Conducting a pilot study and revising your research plan.
9. Selecting the sample to be tested.
10. Collecting the data.
11. Preparing the data for analysis.
12. Analyzing the data.
13. Interpreting the results.
14. Sharing the findings with others.

Of course, the steps you follow will vary according to the kind of assignment you are doing. For example, you would omit some of these steps if you were writing a research paper rather than carrying out an experiment.

Ideally, as a researcher, you should be entirely neutral and objective with respect to the subject matter at hand. However, this is not always as simple as it may seem. Let's say that you are a university scientist and that a major pharmaceutical company funds all of your research. You are asked by the editor of a prestigious medical journal to write a literature review on the effectiveness of a particular drug that is made by your sponsor. How do you go about selecting the studies to include in the review? How do you interpret the results of the experiments that you describe? While you may wish to be as objective as you can, is it possible that you might be subject to some unconscious bias?

Sometimes, even though you have been completely objective and neutral, questions may be raised about the validity of your work simply because you are associated with the industry. It's not difficult to find examples of biased research in controversial situations such this, but you may find that even someone who is doing basic research may be more willing to accept a result that is consistent with his or her hypothesis than one that seems to refute it. The best way to avoid bias is to make your research procedures as rigorous as possible so that your conclusions are unavoidable.

OBTAINING INFORMATION FOR A RESEARCH PAPER

One of the most important rules in writing a scientific paper is that everything you say must be supported by documentary evidence. This rule applies whether you're writing a lab report for a classroom project or a review paper for publication in a professional journal. If you're going to make a career as a scientist, you will find out quickly that you must be able to back up your arguments with facts. When you are just starting out in university, your assignments will probably not require extensive research. However, once you've learned how to track down information efficiently, you'll find that this skill is a powerful asset in many different situations.

WHERE TO START

Let's assume that you have already selected—or been assigned—a topic to write about. (More details about planning a paper and finding a topic are given in Chapters 6 and 7.) Your first task is to find out something about the topic and the names of the major authors in your field. Once you have this information, you can begin to look more systematically for relevant papers by these and other authors. If you have absolutely no knowledge about your topic, perhaps the best way to start is to take the shotgun approach: look everywhere, but do it superficially.

There are quite a few ways to get at least a general idea of what is going on in a particular subject area. One way to do this is to ask around. Many students neglect to take advantage of their principal resource: their course instructor. Although you should not expect that he or she will provide you with all the information you will need, your instructor, or a graduate teaching assistant if you have one, should be able to give you some names and references to get you started. Similarly, if you have friends who have taken the course before or who may be more knowledgeable than you about a particular topic, don't be afraid to ask for their advice on how to approach your assignment.

Armed with this information you can begin your search. At first, you should be seeking general information about your topic so that you can become more familiar with the names of some of the authors in the field and get a sense of the important issues. This used to mean going to the library to look for books and articles on your topic. Now, almost certainly your first step will be to boot up your computer. Even if you want to use the library catalogue, it's very likely that you will have online access. Later in this chapter we will offer some advice on becoming familiar with your college or university library. First, though, we will consider the Internet as a tool for scientific research and offer some suggestions on how to use it most effectively.

THE INTERNET AS A RESEARCH TOOL

When we prepared the first edition of this book in 1986, the Internet barely existed; today, the World Wide Web is the natural place to start when you begin a research project. Further, you have probably found that at least some of your instructors are putting course materials online. This may range from a few lecture notes and some additional information, to a complete course delivered via the Internet. The amount of material available online will continue to increase, so it is essential that you become comfortable working in this electronic environment. You will also need to know about some of the pitfalls associated with online information.

How To Search Online

No single search engine will provide complete coverage. Speed, database size, relevancy rankings, search results, features, and commands vary from engine to engine, so you will probably want to use several engines in the course of your search. Even if the same key words are used as you go from one search engine to another, different results will appear. You should be aware that some Internet service providers might limit the range of sites you visit. For example, some universities now block access to sites from which students might download copyrighted materials, and other providers may block on the basis of content. The following points cover some common essentials:

- **Use lower-case letters when typing keywords**. Upper-case letters may restrict your search; if you use lower-case, the search engine will consider sites that use both upper- and lower-case letters.
- **Use an advanced search to gain access to the most pertinent sites**. Most search engines permit you to do "advanced" searches. They make it easy for you to do this type of search by allowing you to select certain

search limitations, such as language, period in time, or types of occurrence. You can also use *Boolean* operators (i.e., AND, OR, and NOT) to restrict your terms.

- **Try alternate spelling**. To expand a search, try variant spellings of certain words. For example, if you were researching the effects of wall colour on levels of concentration, you would want to search using both "colour" and "color".

- **Save particularly useful websites**. When you find a website you plan to visit again, add it to your "Favorites" list (or, if you are using Netscape or Firefox, your "Bookmarks"). You can then easily return to the site without having to repeat the search procedure.

- **Record the full URL and the date the website was accessed**. You will need this information for citations within your writing.

- **Know when not to use the Internet**. Remember that not everything is available online. A great deal of information can be found only in printed sources. When in doubt, seek guidance from your instructor and reference librarians.

SEARCH STRATEGIES

When you are doing an online search, you want to strike a balance between not missing any relevant information and being overwhelmed with lists of sites that have nothing to do with your topic. Try this simple experiment: imagine you were writing a review paper on the visual abilities of different species of animal, one of which is the Australian marsupial rat. Go to one of the search engines and type in "marsupial." You will see that you that you get a huge list of hits. Now try typing in: "marsupial AND rat AND visual AND acuity." This time the list is much more manageable, and you will see that most of the hits seem to refer to the same topic. What you have done here is used the rules of Boolean logic, which allow you to combine search terms using the logical operators OR, AND, and NOT. You should be aware, however, that the different search engines take different approaches to the Boolean terms. Google, for example, assumes that if you enter more than one word, they all have to be in the page, whereas other search engines will look for sites that contain any of the keywords you enter. Click on the *advanced search* link to get more advice for a particular search engine.

When you plan a search, you should think about the set of keywords and exclusionary words that will lead you to the information that you want and that will keep your search manageable. However, it's probably a good idea to cast your net wide to begin, look at the kinds of hits that you are getting, then refine your search to get exactly what you want.

EVALUATING WEB SOURCES

Later in this chapter we will describe some of the databases that allow you to search for relevant references. Most of these will be journal articles that have been reviewed by professional scientists before they were accepted for publication. If you do a general search online, the situation is a bit different. Most online material is not peer-reviewed, which means that you must be extra cautious about accepting the accuracy or the bias of what you find. The first thing you should do when you come across a website that has information relevant to your research topic is consider the source. Is the site credible? Was it created by an academic institution? a recognized research centre? a group with no professional affiliation? an individual?

Let's say that you have been asked to write a paper evaluating the contribution of animal research to medical discoveries. In your Internet search you find a site belonging to the Research Defence League, which strongly defends the use of animals in research and the contributions they have made. You also come across the website of PETA—People for the Ethical Treatment of Animals—an animal rights organization that vociferously opposes the use of animals in medical research and denies that animal research has made any positive contribution. Both of these organizations have clear credentials, and you understand that what they say must be interpreted in the context of what they believe.

Now suppose you come across another site representing Physicians for Responsible Medicine. The description of this group suggests that it is an independent body that has evaluated the literature and concluded that animal research is fundamentally flawed when it is applied to humans. When you read this material, you conclude that it is quite persuasive in its argument that using animals in research is ineffective. But what if you discover that this organization has a number of links with PETA? Would this influence your interpretation of the arguments they present?

You must be very careful about accepting what you find on the Internet at face value. There is certainly a tremendous amount of information out there, most of which may be very useful to your research project. However, it's always best to approach a site you don't know with a fair degree of caution and skepticism, and you should carefully evaluate what you read before you try to incorporate it into one of your papers.

Here are some tips for evaluating websites:

- Look for a statement identifying the site host and giving the author's qualifications and contact information.
- Check the URL. You can get quite a lot of information just by looking at the top-level domain name of the URL. This is the first part of

the web address after the www, as in "Microsoft.com". All of us are familiar with the *.com* that indicates a commercial site, but other sites have different names. Educational institutions in the United States typically end with *.edu*, while universities in the United Kingdom use *.ac*, as in the University of Durham's *www.dur.ac.uk*. Note that the country might also be represented, with *.uk*, for the United Kingdom, *.fr* for France, and *.au* for Australia. Government documents may be stored at a *.gov* site and many non-governmental agencies may use *.org*. Personal sites are often signified by a tilde (~). By looking at a domain name, you may get a sense of the possibility for bias in the material you find.

- Check to see how current the information is. In many cases this may not be crucial, but if you are looking for something that is completely up to date, you should check when the page was posted to the website or when it was last updated. Some pages will have a footnote providing this information, but others may not. Remember that is easy to get space on a server to post a website; it is also very easy to leave a website sitting there with no updates, even though the information has become quite out of date.

- Evaluate the accuracy of the information by checking facts and figures with other sources. Data published on the website should be documented in citations or a bibliography and research methods should be explained.

- Be wary of blogs. Although some companies have official blogs that can offer good advice about subjects like the stock market and real estate, many are simply online diaries published by a rapidly increasing number of people who are expressing personal opinion and nothing more. Using such unverified material can seriously undermine your essay, as many a student has discovered.

- Don't trust everything you read. Many websites will look like—and will claim to be—authentic, reliable sources of information, but that doesn't guarantee that they are infallible. One such example to keep in mind is Wikipedia. Wikepedia (http://en.wikipedia.org) is an online encyclopedia, but what makes it different from traditional encyclopedias is that anyone can contribute to Wikipedia; as a result, it is always changing and evolving. This also means that it is always susceptible to inaccuracies and misinformation. It may function as an informative starting point, but it is a good idea to verify any information from a site such as this with another source.

Finally, if you still have questions about the reliability of the information you're getting from a particular website, ask your instructor about the organization or group the site represents; he or she will probably have some idea of the organization's reputation in the academic community and how legitimate their information is likely to be. If your instructor isn't familiar with the organization or group, it's probably a good sign that the organization isn't well established, and you may be better off not using their information in your work.

USING THE LIBRARY

It's important to become familiar with the library at your institution. Some of the larger universities have more than one library, each specializing in a different field, and you should get to know what their holdings are. This is particularly important for students of psychology because some of the material you need might be located in the science and medical library, some in the education library, and some in the arts and social sciences library. If your college or university offers library orientation tours, it's well worth taking one. You will learn a great deal about useful reference sources beyond the book and periodical stacks: you may be introduced to bibliographies and guides to literature, dictionaries, encyclopaedias, government documents, and CD-ROM and online databases. Once you know what is available and where the materials are, your literature searches can be much more efficient. You should also take advantage of the knowledge your librarians have; their careers are based on finding and cataloguing materials, and their assistance can be invaluable.

It's also a good idea to make yourself familiar with the cataloguing system your library uses. This will enable you to go to the general area containing materials that may be relevant to your topic. The two main cataloguing systems are the *Library of Congress* system (the most common one in university libraries) and the *Dewey Decimal Classification* system (more common in public libraries). As you can see in the table below, in each system Psychology is broken down into a number of subcategories. So if you are interested in, say, Special Education, you could go directly to the shelves to look for call numbers beginning with "LC" if you have the Library of Congress system, or "37" if you use the Dewey Decimal classification

Table 2.1

	Library of Congress		Dewey Decimal System
BF	Abnormal psychology	00-	Artificial intelligence
	Child psychology	13-	Parapsychology
	Cognition	15-	Abnormal psychology
	Comparative psychology		Child psychology
	Environmental psychology		Cognitive psychology
	Motivation		Comparative psychology
	Parapsychology		Environmental psychology
	Perception		Industrial psychology
	Personality		Motivation
	Physiological psychology		Perception
	Psycholinguistics		Personality
	Psychological statistics		Psychological psychology
HF	Industrial psychology	30-	Family
	Personnel management		Psychology of women
HM	Social psychology		Social psychology
HQ	Family	37-	Educational psychology
	Psychology of women		Special education
LB	Educational psychology	40-	Psycholinguistics
LC	Special education	51-	Statistics
Q	Artificial intelligence	61-	Psychiatry
	Psychological psychology		Psychotherapy
QA	Mathematical psychology	65-	Personnel management
RC	Abnormal psychology		
	Psychiatry		
	Psychotherapy		
T	Personnel management		

USING THE LIBRARY

As we suggested earlier in this chapter, one of the ways to get an overview of a topic is to browse through some of the primary and secondary sources located in the library. Later we will discuss the differences between primary and secondary sources, but for now you should be aware that a primary source is one written by the person doing the study, and a secondary source is one that describes the work second hand. Let's say that you wanted to write an essay on congenital insensitivity to pain. You know that the literature on pain is vast, and you also know that there is relatively little information on congenital insensitivity. You might try the following approach:

1. Have a look for your general topic, "pain," in your library's online catalogue or card catalogue files. You might find that "pain" as a general heading yields too many results, in which case you could try to narrow your search by looking at terms such as "pain" and "insensitivity" together, or else look for "hypoalgesia" or "analgesia." A quick glance will probably show you that most of the books on pain have similar call numbers.

2. Once you find the appropriate stack, browse through some of the books that are there. You don't have to look for specific titles: just go through the indexes of different books for any reference to pain insensitivity. Even if you don't find a full book on your specific subject, you should find a few relevant references; you'll also begin to see how different authors have discussed the topic.

3. Another place to look is in the specialized journals dealing with your topic. Much of the research on pain is published in the journal *Pain*. If you go to the current periodicals section of your library and skim through the contents pages of all of the issues for the past year or so, you might find something useful. This technique doesn't always work, but when it does it can give you a lot of information. Not only will you have this up-to-date reference on your topic, but the reference list at the end of the paper should provide other recent references.

SYSTEMATIC SEARCHING

The strategies outlined above will help you get some basic background information on your topic. If you are writing a review of a body of literature, however, you will need to obtain a much more extensive list of references. You can develop your list of references in a variety of ways, either by examining relevant secondary sources or by searching through the various databases and indexes that are available, either physically within the library or online.

SECONDARY SOURCES

If someone has written a review paper or chapter on your topic in the last few years, it may provide a listing of the earlier literature on your topic; then you can concentrate on tracking down more recent references. You can often find references to review papers by checking your textbooks or by asking your professor. If these approaches fail, you may still be able to get some leads in the library.

One of the best sources for review papers is the *Index of Scientific Reviews*. This is a serial publication listing review articles and chapters that have appeared in the recent scientific literature. Sources are filed under different subject headings and are extensively cross-referenced, so if you are not sure of the best way to describe what you are looking for, you can find alternative headings here.

Another useful source is the *Annual Review* series. These are books of reviews that are published each year and provide detailed summaries of current research. The series includes the *Annual Review of Psychology*, the *Annual Review of Neuroscience*, the *Annual Review of Physiology*, the *Annual Review of Genetics*, and the *Annual Review of Entomology*. These volumes are particularly helpful because they cover most of the sub-areas of a discipline and commission new reviews every few years.

In psychology there are a number of publications that will help you begin your search, whether for topics or for references. If you are looking for a topic to write on, you could consult a handbook, such as the *Handbook of Social Psychology* (Delamater, 2003), *Stevens' Handbook of Experimental Psychology* (Pashler, 2001), or the *Handbook of Psychological Assessment* (Groth-Marnat, 1997). Each of these gives a broad overview of the field and may provide you with ideas for topics to write about.

COMPUTER-BASED SEARCHES

Almost every library now has its catalogue in computer form. In addition, many libraries allow remote access to their catalogues from your home computer or laptop. If you have Internet access at home, you don't have to leave the comfort of your room to be able browse through the catalogues. You may even be able to gain access to other library catalogues if you wish; check with your own library system to see what it offers. Such systems offer several advantages:

- searching is fast and efficient;
- you can search for materials by author, by title, by call number, or by subject, which gives you much more flexibility in your search task;
- you can access the catalogues directly from your home or residence room without having to go to the library;
- you can find out whether the book you need is available in one of the university's libraries or whether it has been checked out;

- if the source you're looking for is not available in your own library, you may be able to access the catalogues of other libraries to see if it is available there.

To get the information you need efficiently, you should know a little about the logic underlying database searches. When you are searching the databases and indexes we describe below, you should use the same Boolean logic that we recommended for searching websites. Although each database has its own particular rules, all require that you enter certain relevant keywords for the computer to search. These may be contained in the title or the abstract of the paper, or they may be part of a list of *descriptors* assigned to the paper by an indexer. When you do a search you should use enough keywords to ensure that you do not miss large numbers of references, but not so many that you come up with too much irrelevant material.

For example, suppose you have been assigned a biology paper on the feeding habits of blowflies. You might begin a general search using the terms "blowfly OR *phormia regina*" AND "feeding OR hunger OR eating." This would give you a list of papers whose titles or descriptors contain some reference to blowflies in conjunction with feeding or eating. If this initial search produced too many references for your purposes, or if many of them were not closely related to what you are looking for, you could narrow the search by including more keywords, such as "preference," "taste," or "behavior." (Remember that most of the databases come from the US and use American spellings.) By selecting the appropriate combination of search terms, a skilled searcher can usually come up with a moderate list of useful references. If you are having trouble establishing an appropriate search strategy, be sure to consult one of the reference librarians; they have a lot of expertise in this area.

ABSTRACTS, INDEXES AND FULL-TEXT DATABASES

When you have to start looking for your references from scratch, probably the best approach is to use the abstract or index journals. These may be available in hard-copy versions in the reference section of your library, but you should always check with the librarians to see if they are also available electronically. If they are, you will find that the search procedures are much simpler and faster.

There are a surprising number of abstract and index journals, covering almost every major subdivision of the social and life sciences, from *Aquatic Sciences and Fisheries Abstracts* to *Weed Abstracts*. Among the most useful of these journals are *Biological Abstracts*, *Excerpta Medica*, *Index Medicus*, *Psychological*

Abstracts, and *Social Sciences Index*. Abstract journals contain abstracts of publications that you can search on the basis of keywords or author names. Index journals, such as *Index Medicus*, provide lists of references without an abstract.

Another kind of index that allows you to find references, though in a somewhat different way, is the citation index. Two of these will be especially relevant to you: the *Science Citation Index* (http://scientific.thompson.com/sci/) and the *Social Sciences Citation Index* (http://scientific.thompson.com/ssci/). These indexes let you search forward in time, in a manner of speaking; that is, if you know of an important paper written by a certain author several years ago, you can find out who has referred to that paper since it was written. In what follows we have described the procedure for using the online version of the *Citation Indexes*. However, these indexes are also available in hard-copy versions, so if you do not have access to the online versions, you may be able to find bound copies at your college or university library.

The *Citation Indexes* allow you to do a general search for articles by a particular author or on a particular topic, but they also enable you to search for articles that cite a particular author or work. So, if you know of one paper written by a certain author, you can enter this information and request a listing of later articles that refer to this paper. Although not all of these citing articles may be directly relevant to the topic you are interested in, you will certainly find out about what kind of work has followed this original article. You may also discover the names of other researchers who are working in the same area, and you can then see who has cited their work. In this way you can build up a collection of core references that will allow you to make progressively wider searches, which is especially important if you are preparing the introduction to a thesis. For a short class essay, on the other hand, you can stop your search once you have identified the major papers in an area.

There are now many databases available online that can be accessed by anyone with an Internet connection. Some of these are electronic versions of print-based volumes. For example, *PsycInfo* is the online (and more extensive) version of *Psychological Abstracts*. Some of these databases will provide abstracts only, while others provide full-text versions of journal articles. Typically, these will be in the form of a PDF file, so that you can print a copy of an article in the same form that it was published in the journal. Some of these databases, such as *Entrez PubMed* offer their services without charge, while others require a subscription fee.

Many university and college libraries subscribe to a number of databases so that students and faculty can access them without having to pay a subscription fee. In Ontario, the Ontario Council of University Libraries has developed *Scholars Portal*, which is available through all participating institutions. *Scholars Portal* provides a single point of entry into a variety of scholarly resources, including full-text journals and several of the databases we will describe below. According to a recent survey, over 40 per cent of the users were students doing research for coursework, and almost half of the users were working from off campus. You should check to see if your own university has access to *Scholars Portal*.

Even without *Scholars Portal* your library may give you access to a variety of different databases. In the table below, we have listed several of these with some brief comments on which areas they serve. For those of you in Psychology, *PsycInfo* and the related services *PsycArticles* and *PsycBooks*, will likely be the most valuable. These will give you access to most of the psychological literature, going back many years. In the life sciences, *Entrez PubMed*, run by the National Institutes of Health in the United States, provides extensive coverage of the medical and health literature. The databases we have listed are only a few among many, so you should check with your own reference librarians to see what is available where you are studying.

Academic Search Premier (http://www.epnet.com/)	Full-text database. Coverage includes engineering, computer science, medical science, and social science.
JSTOR (http://www.jstor.org/)	Full text database. Coverage includes both life science and social science journals.
PsycARTICLES (http://www.apa.org/psycarticles/)	Full text database of articles in journals published by the American and Canadian Psychological Associations, among others.
PsychBOOKS (http://www.apa.org/psycbooks/)	Full text database of books and chapters published by the American Psychological Associations, including many historic texts. It also contains all 1,500 entries of the APA's *Encyclopedia of Psychology*.
PsychEXTRA (http://www.apa.org/psycextra/)	Mostly full-text database of the "gray" psychological literature compiled by the APA. Supplements *PsycInfo* with less technical content (magazines, newspapers, government reports, etc.).

PsychINFO (http://www.apa.org/psycinfo)	The online version of APA's *Psychological Abstracts*. Includes every abstract from APA journals back to 1887 and also covers a wide range of books.
ScienceDirect (http://www.sciencedirect.com/science/)	Mostly full-text database containing journals and books published by Elsevier and from other sources.
Web of Science (http://scientific.thomson.com/products/ *wos/*)	Provides online access to the *Science* and *Social Sciences Citation Indexes*). In addition, *Century of Science* is a database of ground-breaking papers published over the last hundred years.

One other resource that is worth looking into is *Google Scholar* (http://scholar.google.com/). *Google Scholar* is part of the Google system, which gives you access to a wide range of scholarly literature, not just in science. In addition to books and journal articles, you can search for theses or for papers that are in preprint repositories. If you need to begin with a very broad search, *Google Scholar* is a good place to start.

REFERENCES

Delamater, J. (Ed.). (2003). *Handbook of social psychology* (4th ed., Vols 1–2). New York: Kluwer Academic/Plenum Publishers.

Groth-Marnat, G. (Ed.). (1997). *Handbook of psychological assessment* (3rd ed.). New York: Wiley & Sons.

Pashler, H.E. (Ed.). (2001). *Stevens' handbook of experimental psychology* (3rd ed., Vols 1–4). New York: Wiley.

Roeder, K.D. (1998). *Nerve cells and insect behaviour*. Cambridge, MA: Harvard University Press.

chapter 3

KEEPING TRACK

Collecting information from various sources and data from experiments is fundamental to the research process. However, there is another essential part of the research effort that students tend to overlook, even though it may make the difference between an average grade and an excellent one. We are referring to the way you organize your research materials.

MAKING NOTES

A large part of your academic career will be spent obtaining information. This information gathering may involve attending lectures or seminars, reading books and journals in the library, or running experiments. The sheer amount of information that you have to digest can be quite overwhelming and will be impossible to deal with unless you take a systematic approach to organizing it. In doing this, there are three steps that you should take:

1. Summarize your material by taking notes;
2. Organize your material in such a way that you can access it easily and arrange it in different orders; and
3. Make sure that you keep track of the sources of your information.

Reading and lecture notes will be useful for many purposes: when reviewing course material for tests; when writing research essays; when creating bibliographies or research proposals; when preparing for discussion groups, seminars, tutorials, or labs; and when studying for final examinations. Since so much depends on the quality of your notes, you should take the task of writing them seriously. Doing a good job of note-taking right from the start will save you hours of agony later.

LECTURE NOTES

When you attend a lecture, you don't have much time to arrange your notes in any systematic order. Most students just jot down the main points in the hope that they can expand on them later. Unfortunately, most students never get around to doing that, and those few scribbled notes are all they have when the time comes to start preparing for exams. On the other hand, some students try to catch everything their lecturer says, and at the end of term have to decipher notes that, though detailed, are badly organized and make little sense. They may have been concentrating so much on recording the lecture word for word that they failed to grasp the main points.

One of the keys to good note-taking is understanding that this process does not begin with the lecturer's opening words and finish with the end of class. Since lecture notes are so important, it's definitely worth taking some time to prepare beforehand and to review afterwards. Try dividing your lecture note-taking into three phases: preparations before the lecture; note-taking during the lecture; and review and revisions after the lecture.

BEFORE THE LECTURE

It's a good idea to plan how you will deal with lectures in each of your courses before classes begin. Here are some tips:

- Keep separate lecture notes for each course, either in a large binder with separators or in smaller, individual binders, one for each course. A binder is more useful than a fixed-page, stapled or spiral-bound notebook because it allows you to insert any handouts alongside your notes.
- Prepare for each lecture before the class begins. You should have read any assigned pre-lecture readings before going to class. This will help you identify the major points and provide a context for the lecture. An additional benefit to completing your reading assignments beforehand is that you will be more attentive to any information given by your lecturer that does not appear in any of the assigned readings.
- Your instructor may provide an outline of the day's lecture, either displayed with an overhead projector or data projector, or given as a handout. If so—and if he or she is faithful to the outline—it will be easy for you to jot down the major points, perhaps a key phrase or two, and any references provided. If there is no outline, or if the lecturer is disorganized, you will have to listen even more attentively. Your own outline will be especially helpful later as you reflect on the purpose and content of the lecture.
- If you are lucky, your instructor may provide you with an online copy of each lecture on PowerPoint. If he or she does, be sure to take advantage of this information. Print out the slides as handouts, with the

slides on one side of the page and space for notes on the other. You can skim through the slides before the lecture so that you have a sense of what will be covered, then take notes to supplement the material on the slides. This will help to make your studying easier at exam time, and if you need to speak with your instructor about something you don't understand, you can go right to the slide that he or she discussed in class.

DURING THE LECTURE

Remember that the lecturer's intention is not to summarize the textbook or other assigned readings but to present some essential points about the topic under consideration. The lecturer may expand on one or more points in the textbook, give additional examples gained from his or her own field or lab research, outline a concept or principle, work through a problem, present and discuss a hypothesis, explain particularly complex problems, and so on. Your task is to listen and then link what you have heard with what you have read so that you develop a comprehensive understanding of the key issues. Here are some things to remember:

- Start your note-taking on the very first day of the course; important organizational information may be given that will be useful later.
- Start the notes for each lecture on a clean sheet of paper, recording the date and title of the lecture at the top of the page. Taking notes will help you focus on the lecture, but don't spend the whole time writing: listen carefully, think as you listen, and try to understand. Some lecturers will make a major point and then expand on it or illustrate it with examples. Be attentive so that you can hear supporting information and insights as well as the main points.
- Write down key words and phrases, names, places, dates, equations, and numbers.
- Write down any definitions given by the instructor, as they may differ from the definitions of the same terms given in the course readings.
- Copy into your notes any diagrams or figures that the instructor presents.
- As the lecture proceeds, be sure to jot down in the left margin any questions you may have about points you don't understand or information you are unsure about (perhaps because the lecturer was unclear or spoke too quickly); that way, you can clarify these points

later on, either at the end of the lecture or during your instructor's office hours.

- Near the end of the lecture, be especially careful to listen for the instructor's conclusions; write them down accurately and concisely.
- If there is a question period after a lecture, listen closely; someone else's question may be just as important as the lecturer's response. Also, don't be afraid to ask a question, even if you think it's too simple—you may find that others share your puzzlement and will be grateful to you for speaking up.
- Leave room at the end of each day's notes so that you can add comments later, after you have reread the notes and reflected on them in relation to other lectures and readings.

AFTER THE LECTURE

You should review your notes after each lecture while the material is still fresh in your mind; this is a good time to make corrections and additions. You should consider typing your notes into your computer as soon as you can, especially if they are messy. Even if you make your notes by hand, by rewriting them very soon after class, you'll be able to include information that is still fresh in your mind, and you won't have to wonder, *What did I mean by that?* when you start studying for your exams. You can add headings, fill in information you didn't have time to write down, and even make additional comments about things you don't understand. By doing this, you will provide yourself with an organized set of study notes that you can relate to your textbook readings. In addition, the very act of writing out your notes will reinforce your understanding of the material and help you remember it later on.

When you are writing up your notes, keep in mind that you may not go back to them for several weeks, or even longer. Don't be too cryptic or too vague when writing out notes that seem clear enough immediately following the lecture but might make less sense later on.

TWO WARNINGS

First, if you miss a lecture, don't ask the instructor to lend you his or her notes. Your instructor's lecture notes are personal and may not be in any form that will make sense to another reader. They will certainly not include the sorts of spontaneous comments that are often highly pertinent. If you must borrow notes, get them from a classmate you trust to take good notes covering everything that has taken place in class.

Second, if you want to record the lecture, you must obtain permission to do so from your instructor before the class. Many professors won't mind if you tape their lectures, but it's never wise to assume this.

RESEARCH NOTES

When taking notes on research you are doing in the library or on the Internet, you will have more time to organize your material than you have during a lecture. Still, you should think carefully about your approach before you begin. Don't just head off to the library with pen and notepad, planning to take as many notes from as many different sources as possible. Instead, think carefully about how you expect to use your notes and what form you want them to take; think about what will be easiest for you later on.

For example, imagine you are writing a research paper. You may decide you want to try several different ways of organizing your materials to see which makes the best sense. For instance, you may want to change the order of the topics that you plan to discuss. If you have made all your notes in a continuous fashion as you read through the various sources, it will be very difficult to arrange them in a different order. It is much better to put each idea or summary statement on a separate sheet of paper so that you are not locked into a particular sequence. Many students use index cards for their preliminary note-taking because these usually have just enough room for one thought, idea, or quotation, and can be rearranged fairly easily. If you prefer to type your notes directly into a computer, then rearranging ideas is almost no trouble at all. Below we have listed a number of points that may help you to improve the quality of your research notes:

- Use an index card or separate sheet of paper for each topic or idea. The more flexibility you can give yourself for rearranging your material, the better.
- Record the citation information about your source on a separate index card. This way you can keep track of all your sources and build up a bibliography as you go along.
- Be scrupulous in recording the details of your source. Be sure to include the full list of authors or editors, the title, and the complete publication details. If your source is a book, include the name of the publisher and the year of publication. You should also include the total number of pages in the book because this is sometimes required for a citation. If you have only referred to a single chapter, then you'll need the title and authors of the chapter as well as the chapter page

numbers. If your source is a journal, include the title, year, and volume number of the journal, and the title and page numbers of the article. It's also a good idea to include the library catalogue number or—if you're using an online journal or other Internet source—the web address, in case you have to go back to that source again.

- For most of your papers you'll be expected to use some standard citation format, such as the one recommended by the American Psychological Association. (For more on this and other citation formats, see Chapter 11.) If this is the case, you can save yourself a lot of time later by transcribing your references in that format so that you can enter them into the reference section of your paper easily.

- Devise some shorthand method of referring to your sources on index card notes. That way, you will know where you found the information. A simple way is to use a regular citation format. So, at the top left or right of each card, you would put, for example, "Ciccone & Zander (2001a)," which would refer you to the citation card containing the complete publication details. Be sure to include the page where you got the information; this is particularly important if you are quoting a passage from the source.

- In general, unless you want to quote a particular passage, you should avoid direct quotes from the source and even close paraphrasing. For one thing, this is an inefficient way of making notes; more importantly, it leaves you open to charges of plagiarism if you copy your notes directly into your paper without realizing how close your paraphrase is to the original. If you do paraphrase, be sure to make a note to yourself that you have done so.

- When you do quote a source, be sure to do so exactly. If a word is spelled incorrectly, leave it that way and follow it with [*sic*] to indicate that the misspelling appears in the original. If you omit part of a sentence you're quoting, show this by using an *ellipsis*—three spaced dots (explained in detail in Chapter 17).

- Even if you type your notes directly into your laptop the same rules apply. Be sure to set up a system where you can keep your notes, quotations, and reference citations separate. This will make it easy to cut and paste when you finally put a paper together.

LAB RESEARCH

Although this is not a book on research methods, there are certain aspects of the research process that lend themselves to a book on "making sense." If you take a course on research methods, you will be taught how to design and run experiments and how to interpret the results. However, most methodology textbooks do not say much about how to keep track of all of the data you collect and how to organize it in a manner that will make it easy for you to do your data analysis or final write-up.

Once you have chosen or been assigned a research topic, the most important thing you can do is ensure that everything about your project is well organized. This means keeping complete and accurate records in a form that is easily accessible. We have provided hints on a few aspects of note-taking for lab reports below.

EXPERIMENTAL PROTOCOL

One of the most important components of a lab report or a journal article is the *Methods* section. This section is the description of the experiment that produced the findings the author is reporting. It should include enough detail that the reader will know exactly what the author did and how, if necessary, to repeat the experiment in all its essential aspects.

Surprisingly, few students, before beginning an experiment, bother to make a summary of what they intend to do. Instead, they try to *recall* the procedure when the time comes to do the write-up. If you're writing a thesis or an article for publication, there might be quite a delay between the time you run the experiment and the time you do the final write-up. Sometimes, if the apparatus has been dismantled since you completed your study, you may have no way of gathering missing information. For this reason, it makes sense to put all this information together as you set up your experiment. The kinds of information you might need include the following:

- your experimental hypothesis;
- your overall research design, including a description of the experimental and control conditions and of the experimental design you plan to use;
- a list of the groups and individuals to be tested under the different experimental conditions, and the number of subjects that will be tested;
- detailed specification of the equipment you will be using, including brand names and model numbers;
- detailed descriptions of your experimental stimuli, where appropriate. For example, if you were using blocks of different colours to test counting ability in young children, you would describe the number of blocks, their shapes, and their sizes, as well as their colours;
- quantitative descriptions of your experimental materials. These might

include levels of light or sound intensity, concentrations of liquids, weight of reward pellets in an animal study, and so on;

- a description of the experimental procedures that you plan to use, including the number of trials;
- proposed data analysis. Here, you would indicate the way in which you plan to analyze your data, including the statistical tests you plan to use. If you want to be really conscientious, you might also note how you would expect these analyses to turn out if your hypothesis was correct;
- any additional information that is specific to your experiment.

Obviously, each study will be somewhat different, but you can easily modify this list to suit your own needs. One way of organizing your protocol information is to put the points in the order you will present them in your write-up.

DATA TABULATION

In some courses, you may be required to keep a lab book in which you record all the information about your experiments. In other courses, however, this may not be the case, and you will be expected to set up your own data tabulation sheets. These are where you will keep all the information on the testing of each subject, as well as all of your raw data, so they are crucially important. Whatever you do, don't simply scribble numbers down on a piece of paper with the intention of transcribing everything in a neater form later. Your data sheets should be planned and made up before you start any of your testing, so that all you have to do is fill in the appropriate blanks.

For every experiment, you should ensure that you have a list of all the conditions relevant to the data that will be included on each data sheet. If you are running a psychology experiment, there should be a notation identifying individual subjects and the order in which they were tested. Also, there are two things to keep in mind when you design your data sheet. First, you should set it up in such a way that it is easy to enter the data that you collect; that is, you should not have to search for the right spot to enter your results, which might increase your chance of making an error. Second, you should consider how you will be entering your data into the analysis or graphing program that you intend to use. It is very frustrating, when you are trying to enter a series of data points into your analysis package, to have to look on separate pages for successive numbers on a list. Try to keep the numbers you will be entering in sequence on the same data sheet.

Your data sheet should contain at least the following:

- the date;
- the title of the experiment;
- the experimental condition(s);
- subject identification (including computer data file name if applicable);
- a simple grid arrangement to allow you to enter the data easily.

If you have prepared your data sheet properly, you should be able to return to it six months later and know exactly what the data represent.

chapter 4

ETHICAL ISSUES IN RESEARCH AND WRITING

Until recently, there was a strong tendency for the average person to think of science as a "pure" endeavour, and of scientists as disinterested individuals in pursuit of nothing but "truth." This is certainly the view of science promoted by those TV commercials in which a white-coated actor points to a set of "clinical studies" as proof of how effective a certain product is. In fact, science is not always driven by idealistic motives, and scientists, like all people, are motivated by a range of interests. Some are concerned above all with their research, and they take great care in the way they go about it and great pride in the belief that the work they're doing will benefit humankind. Others seek primarily wealth and glory, and may be willing to cheat to achieve these material ends.

Outright scientific fraud is uncommon, but it is of sufficient concern that the U.S. National Institutes of Health have a special office devoted to investigating it. Although all major journals require that articles submitted for publication be reviewed by experts in the field before being accepted, *Science* magazine, one of the most prestigious of these, now requires that investigators have their papers reviewed by their own colleagues *before* they submit them for consideration.

For the most part, only the most glaring frauds reach the public through the media—usually when people are discovered to have falsified a significant part of their data. Stories of major scientific fraud emerge from time to time (Decoo, 2002; Judson, 2004). However, most cases of questionable scientific behaviour are more prosaic: for example, a researcher may discard data from subjects who do not appear to be performing in accordance with expectations, or "adjust" a few data points to make the results look more convincing. Although such behaviour is obviously unacceptable, it almost certainly occurs at every level, from first-year undergraduate labs to major research projects that are supported by millions of dollars in grant funds every year.

The purpose of discussing such unethical practices is not to disillusion you but to start you thinking about what is right and wrong when you are working in the sciences, or in any other academic discipline. Later in this chapter we will examine the question of appropriate documentation of your work, but first we'll take a look at what constitutes ethical scientific research and reporting.

ETHICAL ISSUES IN THE CONDUCT OF RESEARCH

In the 1930s, a developmental psychologist named Wayne Dennis, who was interested in the role of experience in human development, took over the care of a set of fraternal twins within a few weeks of their birth (1935; 1938). Dennis (1935) claimed that he did so "because the father failed to provide for them," and he explained that "the mother understood that we offered temporary care of the twins in return for the privilege of studying them" (p. 18). Because he was interested in their motor development, he kept the babies lying on their backs in their cribs and allowed them to be picked up only for feeding and changing. They were given no toys until they were 14 months old and were hand-fed so that they could not practice reaching. In addition, Dennis "kept a straight face in the babies' presence, neither smiling nor frowning, and never played with them, petted them, tickled them, etc." (p. 19).

When we consider Dennis's research from today's perspective we are horrified, for such a study carried out now would certainly lead to charges of child abuse. It is important to remember, though, that this study was done long before anyone was aware of the effects of social deprivation on young children, and Dennis no doubt believed that what he was doing was in the children's best interests. What was also different about the research environment at the time that Dennis was doing his work was that there were no ethical review committees to consider whether such experiments were appropriate. Individuals back then made their own decisions about what experiments to do and whether or not they were ethical. Nowadays, there is much greater scrutiny of experimental proposals.

When psychology students are first asked to carry out independent projects, they occasionally propose to study the effects of drugs, alcohol, or exposure to sexually explicit materials on some aspect of behaviour. Although there is nothing intrinsically wrong with such studies, they do require that the researcher observe a strict code of conduct. Whenever an experiment is carried out using human subjects, two requirements must be fulfilled: first, there must be some justification for doing the study; second, the subjects who participate must be thoroughly informed about the nature of the research they are becoming involved in and what will be required of them.

JUSTIFICATION

At some time you have probably read a newspaper article reporting an unusual scientific finding and have said to yourself, *So what?* In other words, you were questioning whether that research was worth doing. Although most research projects are driven by curiosity, they also tend to have either a theoretical or a practical rationale. It is usually not enough to begin with the question, "I wonder what would happen if . . . ?" For one thing, most journals will not publish papers that do not offer a good rationale. In addition, when you try to write up a study of this kind, it is difficult to find anything useful to say in either your *Introduction* or your *Discussion* section, as these are the parts of your report where you would ordinarily explain your reasons for conducting the research.

Today, all research projects involving humans or animals must be approved before they can be run. At the undergraduate level, the person responsible for approving research projects may be either the course instructor or the chair of the department. In the case of research funded by granting agencies, specific approval must be given by an ethics committee within the university. Depending on the kind of research, one or more different committees may be involved.

In general, the more invasive a study is, the stronger the justification must be. For example, a study to find out whether ingesting the hallucinogenic drug LSD improves a person's perception of colours is less likely to be approved than a study of the effects of minor sleep deprivation on the detection of visual stimuli in a driving simulation test. In addition, most institutions have a policy of not approving undergraduate projects that have any potential to cause harm to research participants.

INFORMED CONSENT AND FEEDBACK

In addition to ensuring that your study is carried out in a way that minimizes the risk of harm to your research subjects, you must always make certain that your participants have a clear idea of what the study is about and that they understand they have the right to refuse to participate, or even to drop out midway through the experiment, if they wish. In almost all cases, the researcher does this by preparing an *Informed Consent* form, which describes the essential aspects of the study. The subjects read the form and sign it to indicate that they understand and accept the terms under which they will participate.

Sometimes, you may want to exclude certain individuals from participating in a study, but in order to see whether the exclusion criteria apply to your potential subjects, you may have to ask them to reveal personal or possibly embarrassing information. For example, if you were running a study in which subjects would be required to drink alcohol, you would not want to test anyone who may be pregnant, or who has a drinking problem, or who has a disease or medical condition that might be exacerbated by alcohol. Although you could ask questions on your *Informed Consent* forms, some people may feel embarrassed about checking off an item about their personal lives. A simple way to get around this problem is to provide a list of all conditions that would automatically exclude someone from the study and to ask that potential participants excuse themselves if any of the conditions applies to them. That way, they can leave the study gracefully without revealing information about themselves unnecessarily.

Typically, details of the experimental hypothesis are withheld from subjects at the beginning of the study to reduce the risk of affecting the outcome of the experiment. Individuals can influence the outcome of the research they are involved in if they know the goal of the research and try to act in a way they think is consistent with this goal instead of behaving naturally. Thus, in some studies it is even necessary to prevent subjects from knowing the main purpose of the experiment because revealing this information would certainly alter the outcome. Such deception may be permitted if the investigator can justify it.

Whether the study involves deception or not, it is essential, once the experimental session is over, that participants be given detailed feedback explaining exactly what the experiment was about. This is usually done by means of a feedback sheet that describes the study and provides references for follow-up. The experimenter should also be prepared to provide participants with further details about the study if requested.

ETHICAL ISSUES IN WRITING

When you carry out any piece of research, it is assumed that you will do it with appropriate regard to all ethical guidelines. Similar ethical standards apply to any written work you submit.

If you look through most university calendars you will find a section dealing with academic offences. These fall into three broad categories: cheating on examinations (which we will not discuss here), fabrication, and plagiarism. Penalties for academic dishonesty can be severe; they range from a reprimand or a failing grade in the course to expulsion from the university or college in the most serious cases. Despite these penalties, however, many students submit work that is not entirely their own. In a recent survey of academic integrity at a major Canadian University an average of 27 per cent of students admitted to having engaged in the following cheating behaviours at least once:

- Unapproved7 collaboration
- Copying a few sentences from a text without footnoting
- Copying a few sentences from Internet sources without footnoting

- Getting questions and answers from previous test-takers
- Falsifying lab data

It is interesting that most students had been informed of the University polices on cheating and so were aware of what was inappropriate and the possible penalties.

Often it is easy to identify material that has been taken from elsewhere. First, the writing style of most students is different from that of the authors they are copying from. Second, instructors are often so familiar with the sources used by students that they will easily recognize where the material has been lifted from. In other cases, even if the writing style is not distinctive, the quality of the logic and argument may be well above what the instructor expects from a particular student. The following are some examples of what would be considered academic fraud on the part of a student.

FABRICATION

Fabrication occurs when a student inserts false information into a paper. You will be fabricating information if you:

- invent data from an experiment you were supposed to conduct;
- pad a reference list or bibliography with sources you did not use in preparing the paper; or
- include information that did not come from the cited source.

In each of these cases you are trying to gain credit for work you did not do. Submission of falsified data is treated as outright fraud. Even adding an extra reference here and there compromises the integrity of your work and is considered an academic offence.

PLAGIARISM

Plagiarism occurs when you present someone else's words, ideas, or data as your own. You are plagiarizing not only if you use direct quotations without attributing them to their source but if you follow the structure and organization of someone else's work—that is, if you copy a theme, paragraph by paragraph, from someone else, even if you use completely different words. You should also be aware that if you follow a source too closely, even if you acknowledge it, you may still be regarded as having copied from that person. Your work must be a product of your own thought processes, not just a minor modification of something you have read.

Examples of plagiarism range from paraphrasing a sentence or two without appropriate acknowledgement of the source to submitting a paper that has been lifted wholesale from somewhere else. Discussions with students who have been accused of plagiarism reveal that many do not realize that what they have done is inappropriate. This results in what could be termed "inadvertent" plagiarism.

A common cause of inadvertent plagiarism is the failure to record adequate information while taking notes from source materials. For example, you might have jotted down a series of direct quotations from a variety of sources while researching your topic but failed to place them in quotation marks. Later, when writing up your essay, you may string together sentences and phrases from these research notes, forgetting that they are direct quotes. This kind of inadvertent plagiarism can easily happen if you have been looking at material from the Internet and using the cut and paste functions to create your working notes. When this is the case, it's a good idea to include a reminder to yourself that this is what you have done. Even though you may not have intended to do anything dishonest, this is still plagiarism. At this stage of your academic career, ignorance is not a valid excuse: it is up to you to know the difference between original research and copying someone else's words and ideas.

Another situation that sometimes gives rise to accusations of plagiarism occurs when students have been asked to work on a project together but to do independent write-ups. In such cases students may copy directly from each other and submit reports that contain identical passages; or they may do the final write-up separately, but from an outline that they have prepared together. The result of the latter scenario is two papers that are written in different words but with identical organization and structure. This, too, would be considered plagiarism. When you write a paper independently, all the creative aspects—not just the actual words—must be your own.

To give you a clear idea of what might be considered plagiarism, here are some examples, all based on the same original source, with some comments on why the student version is inappropriate.[1]

Source material:

> Why would scientists want to study the brains of other animal species, if their ultimate goal is to learn about *human* thought and behavior? Scientists interested in how the nervous system controls behavior study other kinds of animals for three main reasons (Bullock, 1984). First, they do so to understand the evolutionary history, or phylogenetic roots, of the human brain. They trace what is old and what is new in the human brain—what evolution has brought about.

Second, they try to discover general rules or principles of brain function. Scientists who take this approach ask two different kinds of questions: (1) What in the nervous system correlates with known behavioral differences among animals? For example, if one species is aggressive and another is passive, what differences in their brains account for the difference? (2) What kinds of behavior correlate with known differences in the brains of animals? For example, if a certain brain structure is present in humans but not in other primates, or if a brain structure is larger in one species than another, how do these differences relate to behavior?

Third, scientists study the nervous systems of other animals to obtain information that is impossible to obtain from humans. Many studies that, for technical or ethical reasons, cannot be carried out in humans can be conducted in other animals. Animals provide scientists with "model systems" in which to address questions about how the human nervous system works because the nervous systems of humans and other animals are so much alike. . . . (from Spear, Penrod, & Baker, 1988, pp. 29–30)

Direct quotation:

Why would researchers study the brains of lower animals, if their ultimate goal is to learn about human thought and behavior? Scientists interested in how the nervous system controls behavior study other kinds of animals for three main reasons (Bullock, 1984). First, they do so to understand the evolutionary history of the human brain, tracing what is old and what is new in the human brain. Second, they try to discover general rules or principles of brain function. Third, they study the nervous systems of other animals to obtain information that is impossible to obtain from humans. Many studies that cannot be carried out in humans, for technical or ethical reasons, can be conducted in other animals.

This is a clear example of plagiarism, in which the student has copied the source material almost verbatim, without any acknowledgement. The few minor changes in wording and the omission of several phrases and sentences from the original may have been intended to disguise the fact that this material was stolen.

Moreover, even if this passage were a paraphrase rather than a direct quote, the reference to Bullock (1984) would be inappropriate because the student has evidently not drawn directly from this source but has taken it from Spear et al. (1988). Rather than relying on secondary sources in this way, it is best to go to the original sources. The second-best solution would be to acknowledge the source as follows: (Bullock, 1984, as cited in Spear, Penrod, & Baker, 1988, pp. 29–30).

Paraphrase:

> In my view, there are three main reasons why scientists might study the brains of lower animals in order to learn about human brain function. First, by comparing the brains of animals at various levels of evolutionary development with those of humans, one can examine the evolutionary history of the human brain. Second, general rules or principles of brain function can be ascertained by examining ways in which behavioural differences between species are correlated with differences in their brain structures. Third, using lower animals for research allows scientists to conduct experiments that might be unethical if done on humans.

Even though the wording here is quite different from that of the source material, it still follows the train of thought of the original exactly. In failing to cite the source of these ideas, the student is making a false claim that they are his or her own. This dishonesty is compounded by beginning the paragraph, "In my view," which leads the reader to believe that the ideas expressed by the student are original.

Partial paraphrase:

> Why would scientists want to study the brains of other animal species, if they ultimately wish to learn about human thought and behaviour? As Spear, Penrod, and Baker (1988) have pointed out, there are three main reasons for studying lower animals. First, scientists do so "to understand the evolutionary history, or phylogenetic roots, of the human brain." Second, this type of research allows for an examination of the general rules and principles of brain function, by examining ways in which behavioural differences between species are correlated with differences in their brain

structures. Third, using lower animals for research allows scientists to conduct experiments that might be unethical if done on humans.

Here the student does cite the source of the material but does not adequately acknowledge the extent of his or her debt. Although part of one sentence is placed in quotation marks, several other sentences and phrases that are direct quotes from the original have not been placed in quotation marks. A paraphrase must be entirely the words of the writer; any borrowed words or phrases must be placed in quotation marks. In addition, the student has failed to point out that Spear, Penrod, and Baker drew from Bullock in making these points.

We hope that these examples will cause you to think about how you use your sources. Obviously no one expects you to come up with something that is entirely original and completely removed from what other people have said. The trick is to acknowledge the ideas of others and to use the information they provide as the basis for your own comments. Generally, you don't need to give credit for anything that's common knowledge. For example, if you were discussing the receptive fields of neurons, you would not need to cite the original descriptions because the term is common currency in physiology and psychology. However, if you were discussing specific characteristics of receptive fields, then you would need to refer to your sources. Always document any fact or claim that is unfamiliar or open to question.

Don't be afraid that your work will seem weaker if you acknowledge the ideas of others. On the contrary, it will be all the more convincing: serious academic treatises are almost always built on the work of preceding scholars. If you are unsure whether you are relying too much on your sources, check with your instructor *before* you write your paper.

SUBMITTING WORK THAT IS NOT YOUR OWN, OR THAT HAS BEEN SUBMITTED PREVIOUSLY

No doubt you are familiar with stories about fraternities and sororities that maintain a database of term papers that members can draw upon when they are given assignments. It has always been difficult to determine how much truth there is to such stories, but there is now a new source of material that definitely exists, and it is becoming a serious concern for instructors. An increasing number of websites offering papers for sale have appeared. In some cases these may be drawn from a list of papers that have been written previously and can be copied verbatim. At some of the more sophisticated websites, students can ask for a custom-written essay on a particular topic; these essays may be written by other students who are paid for this work.

Unlike essays that include unattributed quotes that an instructor might recognize, papers that have been purchased over the Internet are much more difficult to identify. However, there are now a number of companies that offer Internet checking of suspect papers. For a fee, universities and colleges can subscribe to a service that will take a paper and compare it with material on hundreds of different websites. A number of universities now have regulations that students could be asked to submit their work in electronic form so that it could be subject to such checking.

One of the most widely used plagiarism detection programs is *turnitin.com*. *Turnitin* is commercially available software that contains a database of millions of sources, including papers that have been submitted previously. The heart of the software is a comparison tool that will assess the similarity between a paper that has been submitted and the information contained in the database. The output is an "Originality Report" that provides the instructor with a side-by side comparison of the student's work and the material found on the Internet. *Turnitin* does not indicate whether a paper has been plagiarized, but it gives the instructor the data to make an informed decision about whether the student has been copying material inappropriately.

Computerized evaluation software is also being used more often to detect cheating in multiple-choice and similar kinds of exams. This kind of software analyzes the patterns of right and wrong answers in tests and exams, and calculates the probability of any degree of similarity. Obviously there will be some overlap in the way people answer questions, but the more right *and* wrong answers that are the same, the less likely it is that they could have occurred by chance. Such coincidental patterns of responses do not prove that cheating has occurred; however, if further investigation shows that two individuals were sitting very close together in the examination and that they happen to be roommates, this could be grounds for concluding that they had committed an academic offence.

NOTE

1   This example is based on a description of academic fraud and the honour system at the University of Virginia, but uses different examples.

REFERENCES

Decoo, W. (2002). *Crisis on campus: confronting academic misconduct*. Cambridge, MA: MIT Press.

Dennis, W. (1935). The effect of restricted practice upon the reaching, sitting, and standing of two infants. *Journal of Genetic Psychology*, 47, 17–32.

Dennis, W. (1938). Infant development under conditions of restricted practice and of minimal social stimulation: a preliminary report. *Journal of Genetic Psychology*, 53, 149–58.

Judson, H.F. (2004) *The great betrayal: fraud in science*. Orlando, FL: Harcourt.

Spear, P.D., Penrod, S.D., & Baker, T.B. (1988). *Psychology: Perspectives on behavior*. New York: Wiley.

chApter 5

WRITING A REPORT ON A BOOK OR AN ARTICLE

In the sciences, you may not often be called upon to write a report on a book. Nevertheless, it is a useful skill to acquire because it prepares you for other kinds of critical writing. The term "book report" covers a variety of writing assignments, from a simple summary of a book's contents to a sophisticated literary review. In between is the kind that you will most often be asked to produce: an analytic report containing some evaluation. The following guidelines cover the three basic kinds of reports. Before you begin your assignment, be sure to check with your instructor to find out exactly which type is expected.

THE INFORMATIVE BOOK REPORT OR SUMMARY

The purpose of an informative book report is to summarize a book briefly and coherently. It is not intended to be evaluative: that is, it does not say anything about your reaction to the work. It simply records, as accurately as possible, in as few words as possible, your understanding of what the author has written. It's not a complicated task, but it does call on your ability to get to the heart of things—to separate what is important from what is not. Aside from some pertinent publication information, all a simple informative report needs to contain is an accurate summary of the book's contents.

READING THE BOOK

DETERMINE THE AUTHOR'S PURPOSE
An author writes a book for a reason. Usually it is to cast some new light on a subject, to propose a new theory, or to bring together existing knowledge in a field. Whatever the purpose, you have to discover it if you want to understand what guided the author's selection and arrangement of material. The best way to find out what the author intends to do is to check the table of contents, preface, and introduction.

SKIM-READ THE BOOK FIRST

A quick overview of the book's contents will show you what the author considers most important and what kind of evidence he or she presents. The details will be much more understandable once you know where the book as a whole is going.

REREAD CAREFULLY AND TAKE NOTES

A second, more thorough reading will be the basis of your note-taking. Since you have already determined the relative importance that the author gives to various ideas, you can be selective and avoid getting bogged down in less important details. Just be sure that you don't neglect any crucial passages or controversial claims.

When taking notes, try to condense the ideas. Don't take them down word for word and don't simply paraphrase them. You will have a much firmer grasp of the material if you resist the temptation to quote: force yourself to interpret and summarize. This approach will also help you make your report concise. Remember: you want to be brief as well as clear. Condensing the material as you take notes will ensure that your report is a true summary, not just a string of quotations or paraphrases.

WRITING THE REPORT

IDENTIFY PRIMARY AND SECONDARY IDEAS

When you're writing your report, give the same relative emphasis to each area that the author does. Don't just list the topics in the book or the conclusions reached; discriminate between primary ideas and secondary ones.

FOLLOW THE BOOK'S ORDER OF PRESENTATION

A simple summary doesn't have to address the topics in the same order in which they are presented in the book, but it's usually safer to follow the author's lead. That way your summary will be a clear reflection of the original.

FOLLOW THE LOGICAL CHAIN OF THE ARGUMENTS

Don't leave any confusing holes. You won't be able to cover every detail, of course, but you must make sure to trace all the author's main arguments in such a way that they make sense. Remember to include the evidence the author uses to support his or her arguments; without some supporting details, your reader will have no way of assessing the strength of the author's conclusions.

TAILOR THE LENGTH TO FIT YOUR NEEDS

A summary can be any length, from one page to six or seven. It depends less on the length of the original material than on your purpose. If the report is an assignment, find out how long your instructor wants it to be. If it's for personal reference only, you will have to decide how much detail you want to have on hand.

READ AND REVISE YOUR REPORT TO MAKE SURE IT'S COHERENT

Summaries can often seem choppy or disconnected because so much of the original is left out. Use linking words and phrases (see Chapter 14) to help create a flow and give your writing a sense of logical development. Careful paragraph division will also help to frame the various sections of the summary. If the report is for a science or social science course, you can probably use headings as well to identify sections.

You may find that you have to edit your work a number of times to eliminate unnecessary words and get your report down to the required length. Editing is a difficult task, but it becomes easier with practice.

INCLUDE PUBLICATION DETAILS

Details about the book (publisher, place and date of publication, and number of pages) must appear somewhere in your report, whether at the beginning or at the end, separated from your discussion by a triple space. Follow the guidelines in Chapter 11 for presenting these details in the correct manner.

THE ANALYTIC BOOK REPORT

An analytic book report—sometimes called a book review—not only summarizes the main ideas in a book but at the same time evaluates them. It is best to begin with an introduction, then follow with your summary and evaluation. Publication details are usually listed at the beginning but can also be placed at the end.

INTRODUCTION

In your introduction you should provide all the background information necessary for a reader who is not familiar with the book. Here are some of the questions you might consider:

- What is the book about?
- What is the author's purpose? What kind of audience is he or she

writing for? How is the topic limited? Is the central theme or argument
stated or only implied?

- What documentation is provided to support the central theme or argument? Is the documentation sound? Is it presented clearly, and does
  the discussion develop logically? Does the documentation support the
  author's contentions and conclusions?
- Is the title pertinent and useful as a guide to the book's contents? How
  has the author divided the book into chapters? Do the chapter titles
  accurately reflect each chapter's contents?
- How does this book relate to others in the same field? To other works
  in the same area?
- What are the author's background and reputation? What other books
  or articles has he or she written?
- Are there any special circumstances connected with the writing of this
  book? For example, was it written with the co-operation of particular
  scholars or institutions? Does the subject have particular significance
  for the author?
- What sources has the author used?

Not all of these questions will apply to every book, but an introduction that
answers some of them will put your reader in a much better position to appre-
ciate what you have to say in your evaluation.

SUMMARY

You cannot analyze a book without discussing its contents. The basic steps are
the same as those outlined above for a simple book summary. You may choose
to present a condensed version of the book's contents as a separate section, followed by your evaluation; or you may prefer to integrate the two, assessing the
author's arguments as you present them.

EVALUATION

In evaluating the book, you will want to consider some of the following
questions:

- How is the book organized? Are the divisions valid? Does the author
  focus too much on some areas or give short shrift to others? Has anything been left out?
- What kind of assumptions does the author make in presenting the
  material? Are they stated or implied? Are they valid?

- Does the author accomplish what he or she sets out to do? Does the author's position change in the course of the book? Are there any contradictions or weak spots in the arguments? Does the author recognize those weaknesses or omissions?
- What kind of evidence is presented to support the author's ideas? Is it reliable and up to date? Are any of the data distorted or misinterpreted? Could the same evidence be used to support a different case? Does the book leave out any important evidence that might weaken the author's case? Is the author's position convincing?
- Does the author agree or disagree with other writers who have dealt with the same material or problem? In what respects?
- Is the book clearly written and interesting to read? Is the writing repetitious? Too detailed? Not detailed enough? Is the style clear? Or is it plodding, "jargonish," or flippant?
- Does the book raise issues that need further exploration? Does it present any challenges or leave unfinished business for the author or other scholars to pick up?
- If the book has an index, how good is it? Are there illustrations? Are they helpful? Is there a list of "further readings" or a bibliography?
- To what extent would you recommend this book? What effect has it had on you?

Remember that your job is not only to analyze the contents of the book but to indicate its strengths and weaknesses. Also, be sure that you review the book the author actually wrote, not the one you wish he or she had written. In short, be fair.

THE JOURNAL ARTICLE REPORT

Many of the suggestions we have made for a book report also apply when you're asked to summarize a journal article, but there are a few additional things you should keep in mind. The purpose of an article report is to show your instructor that you understand what the author has done and quite often to provide a critique. So typically, you should consider the report in two parts. The first part will be a summary that provides an overview of the background, a clear statement of the experimental hypothesis, and a description of the procedures and results. In describing the methods and the results, you do not need to provide as much detail as the author, but you should give enough information that a reader will know all of the procedural steps and the main features of the analysis and the data.

The second component of the report, which is more difficult, is to provide a critique. This does not mean that you should be trying to trash the paper; after all, it would have been subject to an independent review before it was published. But there are almost always ways in which a study could be improved. For example, you might ask if the experimental design was the most appropriate—if the author used an independent groups design, would it have been better to use a within-subjects design? If the results are only marginally significant would it have been helpful to test more participants? Ask yourself the question: "If I was going to run this study again myself, is there anything I would have done differently?" Of course, you should also be on the lookout for flaws in logic, particularly in the Discussion section. Do the conclusions follow inevitably from the data, or is the author speculating too much? Do you feel that the data fully support the stated hypothesis?

As you read more articles you will find that you become more adept at picking out problems, but the important thing to remember is that the task of the author is to present a case, and your job is to decide whether that case has been made convincingly.

cHApter 6

Writing an Essay or Research Paper

Unlike students in the humanities, science students don't always get a lot of practice in writing essays. It is not unusual for students in upper-level psychology or biology courses to admit that they have never written an essay in university. If you are one of the many students who dread writing an academic essay or research paper, you will find that following a few simple steps in planning and organizing will make the task easier—and the result better.

A scientific essay is usually a review of a body of literature, written to support a particular theory, or to outline what is known about a particular topic. The trick to writing a good scientific essay is to organize your material well and provide a good storyline. Try to avoid the kind of paper that is nothing more than an annotated bibliography of the papers you have read: "Solway (2001) studied . . ."; "In 1973, Misener reported . . ."; "An interesting result was obtained by Singh & Rajaratnam (2006)"; and so on. Such papers are not only tedious to read but often difficult to follow. The solution to this problem is to begin with a well-organized outline and stick to it when you actually write the paper.

THE PLANNING STAGE

Some students claim they can write essays without any planning at all. On the rare occasions when they succeed, their writing is usually not as spontaneous as it seems: almost certainly, they have thought or talked a good deal about the subject in advance and have come to the task with some ready-made ideas. More often, students who try to write a lengthy paper without planning just end up frustrated. They get stuck in the middle and don't know how to finish, or they suddenly realize that they are drifting off in the wrong direction.

In contrast, most writers say that the planning stage is the most important part of the whole writing process. Certainly the evidence shows that poor planning usually leads to disorganized writing. For the majority of student papers, the single greatest improvement would not be better research but better organization.

This insistence on planning doesn't rule out exploratory writing. Many people find that the act of writing itself is the best way to generate ideas or overcome writer's block. The hard decisions about organization come after they've put something down on the page. Nevertheless, whether you organize before or after you begin to write, at some point you need to make a plan.

FINDING A TOPIC

Whether the subject you start with is one that has been assigned or suggested by your instructor or is one that you have chosen yourself, it is bound to be too broad for a single essay topic. You will have to analyze your subject in order to find a way of limiting it. The best way to do this is to ask questions about the topic.

Because every essay is different, there are no absolute rules on the questions you should be asking. However, questions about the *context* and the *components* of the situation you are discussing are relevant to almost every essay or research paper:

CONTEXT:

- What is the larger issue surrounding the topic?
- What tradition or school of thought is relevant to the topic?
- How is the topic similar to, and different from, related topics?

COMPONENTS:

- What parts or categories can be used to break down the topic?
- Can the main categories be subdivided?

When you ask about the *context*, you are trying to develop a conceptual framework into which you might fit the whole essay. So, if you are writing an essay on a topic relating to mental illness, one of the first things to ask is where your topic will fit with respect to the various theories about etiology and treatment. Then you can decide how to approach it.

When you ask about *components* and how the topic might be broken down into different categories, you are forcing yourself to analyze what the structure of your paper might eventually be. In other words, this should start you thinking about how you will organize your essay. Suppose, for example, that your assignment was to write about the localization of function in the brain—that is, which parts of the brain are responsible for different kinds of behaviour. You may decide that the main components of this topic are (1) the historical origins in nineteenth-century phrenology; (2) the contributions gained from the study of brain-injured patients; (3) the anatomical and physiological brain mapping experiments that identified the function of particular brain centres; and (4) the recent imaging techniques that allow researchers to study ongoing brain activity while a subject engages in specific tasks. Each one of these components could be the topic of an essay in itself. Once you have broken down the topic in this way, however, you can decide whether to write a brief overview of each of the components, or to examine the way in which the development of technology within these component areas led to the current advances, or to break down one of the components further and discuss its different aspects in more detail.

CHOOSING YOUR OWN TOPIC

Although your instructor will sometimes assign you a topic to write about or give you a specific question to answer, on other occasions your instructor will not restrict you at all, so you will be limited only by the subject matter covered in the course. In such cases, it is important to select a topic that you know you can deal with comfortably. At the same time, remember that the purpose of writing a paper is to learn more about a topic, so the one you choose should not be one that is overly familiar to you. Here are some guidelines for choosing a suitable topic:

- **Choose a topic that interests you.** If your topic doesn't interest you, you won't be able to work up much enthusiasm for reading or writing about it, and your lack of enthusiasm is likely to be reflected in the quality of your work. You will find it easier to spend time and energy working on a topic that interests you, and you will do a better job as a result.

- **Choose a topic on which sufficient material is readily available.** However fascinating a topic may be, you won't get far if your library doesn't have the most important books or journals in that subject area and if you can't find reliable information about it on the Internet.

- **Don't choose a topic that is too difficult.** Although this may seem obvious, students are sometimes drawn to topics that look interesting but turn out to be more technical than they had anticipated. If you decide to write a paper that requires detailed knowledge of a specific field—for instance, neuropsychology or statistics—be sure you have sufficient background knowledge, or you will not understand the literature.

- **Limit your topic to something manageable.** Beware of subject areas that are very broad: you don't want to put yourself in the position of having to deal with too much material. A cursory review of a large subject may appear thin and inadequate to a reader. If you do choose an area with a very large body of literature, limit your scope and be sure to indicate in your introduction that this is your intention.

- **Check with your instructor.** If you have any doubts about whether or not the topic you have selected is appropriate for the course, check with your instructor.

Once you know the general topic you want to write about, the most effective strategy is to look for a distinct subtopic that you can manage in the time and space available to you. For example, under the general topic of "neural plasticity" you would have an enormous number of potential paper topics. To narrow the field, you might begin by drawing a tree diagram of some of these possibilities, as illustrated in Figure 6.1.

Obviously, there are many more options than this, and some of the topic areas may overlap, but such a diagram can help you narrow your choices down to a number you can handle comfortably.

FIGURE 6.1

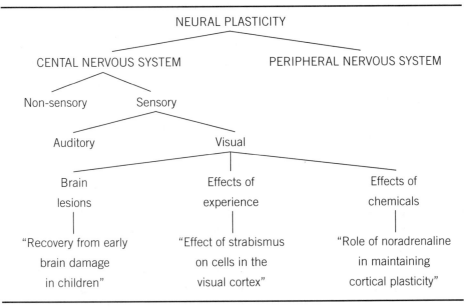

DEALING WITH AN ASSIGNED TOPIC

Even if the topic of your essay is supplied by your instructor, you still need to analyze it carefully. Try underlining key words to make sure that you don't overlook anything that is asked for. Distinguish the main theme of your paper from the subordinate ones. A common error in dealing with prescribed topics is to emphasize one portion while neglecting another. Give each part its proper due, and make sure that you actually do what the instructions tell you to do; if you don't, you may receive a low grade, no matter how good your essay is. For example, *discussing* a subject is not the same as *evaluating* or *tracing* it. Some of the more common instructions are these:

outline	state simply, without much development of each point (unless asked).
trace	review by looking back: examine the stages or steps in a process, or the causes of an occurrence.
explain	show how or why something happens.
discuss	examine or analyze in an orderly way. This instruction allows you considerable freedom, as long as you take into account contrary evidence or ideas.
compare	examine differences as well as similarities. (We discuss comparisons in more detail on page 66–7.)
evaluate	analyze strengths and weaknesses to arrive at an overall assessment of worth.

These and other verbs tell you how to approach your topic; be sure that you know what they mean, and follow them carefully.

PRIMARY AND SECONDARY SOURCES

In Chapter 2 we described how you could find materials relevant to your topic. The next step is selecting the references that you want to read. If you are writing a review paper, or if you are preparing an introduction to an Honours thesis, then it is likely that your instructor will almost certainly expect that your paper is based on primary sources. So, it's important to recognize the difference between primary and secondary sources. A primary source is one written by the individuals who did the original research. Usually this would be a journal article, or in some cases a book length monograph, that describes the work first hand. A secondary source, in contrast, is written by someone who has read the primary material and is providing a summary and comments. In some cases this might be a scholarly review published in a scientific journal, or it may be a book that gives an overview of a wide range of literature. There is also a third kind of source that lies somewhere in between. This is the review and meta-analysis. A meta-analytic paper is one in which the author takes the data from several primary sources and reanalyses the data to help draw more global conclusions. In this case, because it is original work, you should consider meta-analyses as primary sources if you are looking at the original work, but as a secondary source if you are looking at the descriptions of other papers.

In disciplines such as English or philosophy, where students have a good deal of freedom to express personal opinions, instructors may discourage secondary reading because they know that students who turn to commentaries may be so overwhelmed by the weight of authority that they will rely too heavily on them and produce unoriginal, second-hand work.

In the sciences, by contrast, your first task is to find out what is known about a particular topic. Under these circumstances, reading book chapters and review articles, provided they are used judiciously, can be a good way of getting a quick overview of your topic. In fact, it is usually a good idea to read recent literature reviews on the topic you have chosen to learn about the views of the experts in your discipline.

On the other hand, you should not use these reviews as a way around the difficulty of understanding primary sources. Remember that although secondary sources are an important part of research, they can never substitute for your own active reading and analysis of the primary material. Sometimes you will find a review paper whose purpose is to consider the literature in order to support a particular theoretical point of view. In such a case you should be prepared to evaluate whether the author's interpretation of the primary source is the same as your own.

After you have identified the primary materials, however, you should then start reading them as the basis for your paper. The best way to approach these papers is to begin by skimming over them to get an idea of the content. Don't just start reading every article from beginning to end. Read the introductory sections or abstracts of several papers to get a sense of the kinds of questions the authors are asking. Once you have an overview, it will be easier to focus your own questions for a more directed and analytic second reading.

Make no mistake: a superficial reading is not all you need. You will have to work through the material carefully a second time. However, an initial skim followed by a focused second reading will give you a much more thorough understanding of the material than trying to digest it all at once.

ORGANIZING YOUR MATERIAL

We return once more to our theme of organization because this is the foundation of all good work. Once you have started collecting material for your paper, you need to organize it in some way. In particular, there are four important things you need to do:

1. Keep track of where your information comes from;
2. Collect information on different aspects of your topic;
3. Decide on the approach you will take; and,
4. Create an outline that you can follow when you begin to write.

KEEPING TRACK

As we discussed in Chapter 3, nothing is more frustrating than to find that you aren't sure where a piece of information came from. Each time you read an article in a journal or a chapter in a book, write down the reference on an index card. This is important in case you want to find that reference again or in case you need to list it in the *References* section of your completed paper. It may be helpful also to put down the library call number or, if the article comes from an online journal, the Internet address (or URL). Write the full reference in the format that you will use for your reference list (see Chapter 11). As you accumulate more references, sort them into alphabetical order. When the time comes to prepare your *References* section you can simply type them in from the cards. An added advantage of this system is that you can start to build up a database of references that may come in handy for other papers on similar topics.

An alternative to writing index cards by hand would be to type the references into the computer directly, so that you don't have to write them out a second time. A word-processing program such as Microsoft Word will allow you to sort your references into alphabetical order quite easily (use the "Help" function to see how to do this), so that you can enter each reference as you read it, then do a sort at the end when you are preparing your reference list.

If you are likely to be writing a lot of papers, or if you are thinking of going to graduate school, then you might consider purchasing one of the bibliographic citation management software systems. There are several of these on the market, including *EndNote*, *ProCite*, *Reference Manager*, and *RefWorks*. Although the details differ, all allow you to enter a citation for a reference, along with various other pieces of information, such as the library call number, or even the abstract, into a database. After you have created your database of references, all you have to do is to enter a citation code at the appropriate point in your paper for each reference you plan to use. When the paper is complete, you simply ask the pro-gram to create a reference list for you. In most cases the program will not only create the reference list and enter the citations into the text of the paper, it will also format each reference in the style required by the discipline. This is very helpful if you use the same references for papers that use different reference for-mats; you only have to enter the references into the database once, and they will be there the next time you work on a similar topic.

Several of the journal databases we mentioned in Chapter 2 will also allow you to export the citation information for the references that you find directly into the citation management program. You can also share databases with other people who are working in the same area. Finally, a number of University libraries have now made this kind of software available to their users, so you can access it online for free, even if you move from one institution to another. Check to see if your own library provides this service.

COLLECTING INFORMATION

Finding your research material is one thing; making sure that you have correctly summarized what you have read is another. You must take notes that are dependable and easy to use. With time you will develop your own best method, but here again, the index-card system works well either using physical cards or virtual ones on your computer. Using a different set of index cards from the ones containing the full reference, write down the short citation (e.g., Smith & Smith, 1993) so that you can cross-refer to your other cards. Then, below the citation, note the major points of the article. You don't need to go into much detail—just enough to help you remember what the paper was about. (You can always go back if you need more detail.) It's also worthwhile to jot down a reminder of how you felt about the paper. That way, when you're doing your final write-up and you see "useful review" on a card, you can go back and read the article over again; but if you see that it was "incomprehensible" or "too theoretical," you can skip that particular paper.

An alternative to using cards for individual sources, especially if you're famil-iar with the major issues you will be discussing, is to let each card deal with a specific topic. In this case, you would indicate the topic and source at the top of each card and then make your notes. If a single paper deals with several dif-ferent aspects of your subject, you would use several different cards. The advan-tage of this system is that once you have generated an outline, you can arrange your material in a corresponding order for writing.

If you type all of your notes directly into your computer, you should make sure that you cross-refer between your notes and the source so that you can keep track. If you want to shuffle your notes as you would your index cards, there is a simple trick you can use in Microsoft Word. Simply activate the View-Outline function, and each paragraph will appear marked with a bullet. By clicking on and dragging the bullets, you can move paragraphs around very easily in your document.

One final point: whether you use index cards or computer files, be sure not to switch from one system to the other in the middle of your research. It will be much more difficult to retrieve and collate the information if you do.

DECIDING ON AN APPROACH

If your assignment is simply to write an essay on a particular topic, then it is up to you to decide how you will approach it. Whatever approach you take, you must make sure that it will be clear to the reader. You should have a central, controlling idea that will lead the reader through the paper so that he or she will never have to ask, "What is the point that this paper is trying to make?"

There are two general approaches you can take. The first is to organize your essay around a theme that will hold your ideas and information together as you write. This sort of approach is particularly useful for reviewing the literature on a particular subject, discussing recent findings in a particular area, or exploring research trends in a certain discipline. A good strategy for this kind of approach is to write a single-phrase statement of your theme to serve as an anchor before you begin writing; for example, "factors that determine performance on intelligence tests"; or "the development of speech perception in infants." This phrase may become the title of your final paper. Once you have defined your topic in this way, you can then go on to discuss or explain it in whichever way you choose. With this type of approach you are not necessarily taking a point of view but providing the reader with a summary of your understanding of a subject or the available literature on it.

An alternative approach is to develop and defend a particular thesis. This is referred to as the *argumentative form*, and it is usually easier to organize and more likely to produce forceful writing. This approach certainly makes for a more interesting essay, if it is done well. For example, if you were writing on intelligence, you might take the position that cultural bias invalidates the results of many intelligence tests. Or you could argue that intellectual ability is inherited, and that IQ tests are the most valid measurements we have. It doesn't really matter what position you take as long as you make a convincing case for it. Your instructor may not agree with you, but if you provide a well-thought-out argument, you should get credit for it. A note of caution, however: "argumentative" in this context refers to the strategy of presenting reasoned arguments, not one of being aggressive or dismissive of your sources. Good writing is dignified, not belligerent or abusive.

Once you have decided upon one of these two approaches, the next step is to select the single theme or thesis that will serve as the focal point around which you will organize your material. Although you may start with a particular working thesis, it doesn't have to be the final one; sometimes you will change your opinion as you work your way through the literature. A working thesis simply serves as a way to hold together your information and ideas as you organize.

At some point in the writing process you will probably want to make your working thesis into an explicit statement that can appear in your introduction. Even if you don't state it formally, a working thesis will help you define your intentions, make your research more selective, and focus your essay. Therefore, you should take time to develop your working thesis properly. Use a complete sentence to express it, and make sure that it is limited, unified, and exact (McCrimmon, 1976).

MAKE IT LIMITED

A limited thesis is one that is narrow enough to be examined thoroughly in the space you have available. Suppose, for example, that your general subject is sexuality and adolescence. Such a subject is much too broad to be dealt with in an essay of two or three thousand words: you must limit it in some way and create a line of argument for which you can provide adequate supporting evidence. For example, you might want to discuss factors that influence contraceptive use, or the role of TV and movies in setting models for sexual behaviour and attitudes.

MAKE IT UNIFIED

To be unified, your thesis must have one controlling idea. Beware of the double-headed thesis: "Adolescent sexual attitudes in the 1960s were permissive because children were rebelling against their conservative parents, but now their attitudes are more conservative because of concerns about AIDS." What is the controlling idea here? Is it parent-child relationships, or the way that external threats can influence behaviour? It is possible to have two or more related ideas in a thesis, but only if one of them is clearly in control, with all the other ideas subordinate to it; for example, "Although concern over the spread of AIDS resulted in sexual conservatism among adolescents in the 1990s, experts fear this caution is giving way to a growing sense of fatalism and the abandonment of 'safe sex' practices in the early 2000s."

MAKE IT EXACT

It is important, especially when you are defending a position, to avoid vague terms such as "interesting" and "significant," as in "Banting and Best's early failures were significant steps towards their discovery of insulin." Were the failures "significant" because they provided an incentive to work harder, or because they provided insights that could be used later? Remember to be as specific as possible in creating your thesis in order to focus your essay. Don't just make an assertion: give the main reasons for it. Instead of saying, "Intensive planting practices can increase crop yields substantially," and leaving it at that, add an explanation: ". . . because the plants are positioned strategically to allow maximum use of the space available, and the leaf cover provides a living mulch that discourages the growth of weeds." If you are concerned that these details make your thesis sound awkward, don't worry: a working thesis is only a planning device, something to guide the organization of your ideas. You can change the wording of it in your final essay.

CREATING AN OUTLINE

Individual writers differ in their need for a formal plan. Some say they never prepare an outline, while others maintain they can't write without one. Because organization is such a common problem, though, it's a good idea to know how to draw up an effective plan. The exact form it takes will depend on the pattern you are using to develop your ideas—whether you are defining, classifying, or comparing, for example (*see pages 65–7*).

If you find it especially hard to organize your material, your outline should be formal, written in complete sentences. If, on the other hand, your mind is naturally logical, you may find it's enough just to jot down a few words on a scrap of paper. For most students, an informal but well-organized outline in point form is the most useful model. A useful strategy is to begin the outline with single keywords that you can gradually expand into topics and, eventually, complete sentences. With some thought, you can construct your whole paper within the framework of the outline.

If you have used index cards to organize your library research materials according to topic, these can provide a simple way to begin your outline. Rearranging the cards in different orders will give you an idea of how topics fit together before you put the outline down on paper. Nowadays, a more sophisticated approach is to use your computer. Most word-processing packages will allow you to create an outline automatically, with major headings and several levels of subheadings. Using this function, you can move subsections around and change main headings to subheadings or vice versa.

When you are constructing an outline, there are several points to keep in mind. Suppose, for example, you were writing a paper on animal navigation:

I. Introduction
   Thesis: Refined navigational skills are found in many different species, but they serve quite different purposes.

II. Orienting and navigating seen in many different kinds of animal
   A. Birds
      1. Arctic tern
      2. Homing pigeon
   B. Insects
      1. Army ant
      2. Monarch butterfly
   C. Fish
      1. Salmon

III. Different animals use different cues
   A. Celestial cues
      1. Sun compass
      2. Star navigation
   B. Terrestrial cues
      1. Geomagnetism
      2. Barometric pressure
      3. Odour trails
      4. Landmarks

IV. Navigating serves different purposes
   A. Migration
      1. Favourable climate
         a) Going south
         b) Going north
      2. Food availability
      3. Breeding grounds
   B. Locating local food sources
      1. Honeybee
      2. Ant

V. Conclusion

The guidelines for this kind of outline are simple:

- **Arrange your outline according to themes.** In the example above, there are three main themes: 1) the nature of animal migration and the variety of animals that navigate; 2) the cues that different species use; and 3) the reasons they navigate. You could make your first section quite short, simply giving examples of navigational feats in different species. Then, in the second and third sections, you could focus on cues and reasons, again using different species as examples. Arranging your material according to themes will produce a much more readable essay than, say, simply listing various animals and explaining why they navigate and what cues they use.
- **Code your categories.** Use different sets of markings to establish the relative importance of your entries. The example here moves from Roman numerals to uppercase letters to Arabic numbers to lowercase letters, but you could use another system. Most computer outlining programs will provide a default coding system automatically but will also allow you a variety of alternatives.
- **Categorize according to importance.** Make sure that only items of equal value are put in equivalent categories. Give major points more weight than minor ones.
- **Check lines of connection.** Make sure that each of the main categories is directly linked to the central thesis; then see that each sub-category is directly linked to the larger category that contains it.
- **Be consistent.** In arranging your points, be consistent. You may choose to move from the most important point to the least important or vice versa, as long as you follow the same order each time.
- **Use parallel wording.** Phrasing each entry in a similar way makes it easier for the reader to follow your train of thought.
- **Be logical.** In addition to checking for lines of connection and organizational consistency, make sure that the overall development of your work is logical. Does each heading/idea/set of data/discussion flow into the next, leading the reader through your material in the most logical manner?

ONE FINAL WORD

Be prepared to change your outline at any time in the writing process. Your initial outline is not meant to constrain your thinking but to relieve anxiety about where you're heading. A careful outline prevents frustration and dead ends—that "I'm stuck, where can I go from here?" feeling. But since the very act of writing will usually generate new ideas, you should be ready to modify your original plan.

For example, you could alter the outline above by starting off with your discussion of the reasons why animals need to navigate, then looking at possible mechanisms, and finally giving examples of how these cues are used by different species. Sometimes you may start with one sequence and then change it after you have written some of the sections because you can see that it flows better with the new arrangement. Remember that the reason for using an outline is to help you organize your thinking, and that your thinking may change in the course of working on your paper. You should never think of your outline as being written in stone, especially when you have "cut" and "paste" functions available to you. The most important thing is the end product, your paper.

THE WRITING STAGE

WRITING THE FIRST DRAFT

Most writers find it easier to compose the first draft of an essay as quickly as possible and do revisions later than to produce a polished final version in one stroke. However you begin, you shouldn't expect the first draft to be the final copy. Skilled writers know that revising is a necessary part of the writing process, and that the care taken with revisions makes the difference between a mediocre essay and a good one.

You don't need to write all parts of the essay in the same order in which they are to appear in the final copy. In fact, many students find the introduction the hardest part to write. If you face the first blank page with a growing sense of paralysis, try leaving the introduction until later and start with the first idea in your outline. If you feel so intimidated that you haven't even been able to draw up an outline, you might try the approach suggested by John Trimble (1975, p. 11) and charge right ahead with any kind of beginning—even a simple "My first thoughts on this subject are . . .". Instead of surfing the net or running out for a snack, try to get going. Don't worry about grammar or wording: at this stage the object is to get something down on paper or on screen.

Of course, you can't expect this kind of exploratory writing to resemble the first draft that follows an outline. You will probably need to do a great deal more changing and reorganizing, but at least you will have the relief of seeing words on a page to work with. Many experienced writers—and not only those with writer's block—find this the most productive way to proceed.

DEVELOPING YOUR IDEAS: SOME COMMON PATTERNS

The way you develop your ideas will depend on your topic, and topics can vary enormously. Even so, most research papers follow one of a few basic organizational patterns. Here are some of the patterns, along with tips for using them effectively.

DEFINING

Sometimes a whole paper is an extended definition, explaining the meaning of a concept that is complicated, controversial, or simply important to your field of study: for example, the *superego* in psychoanalytic writings, *prions* in microbiology, or the *punctuated equilibrium theory* in evolutionary biology. Rather than make your whole paper an extended definition, you may decide just to begin your paper by defining a key term and then shift to a different organizational pattern. In either case, make your definition exact; it should be broad enough to include all the things that belong in the category but narrow enough to exclude things that don't belong. A good definition builds a kind of verbal fence around a word, herding together all the members of the class and cutting off all outsiders.

For any discussion of a term that goes beyond a bare definition, you should give concrete illustrations or examples. Depending on the nature of your paper, these could vary in length from one or two sentences to several paragraphs or even pages. If you were defining the superego, for instance, you would probably want to discuss at some length the theories of leading psychoanalysts.

In an extended definition, it's also useful to point out the differences between the term in question and other terms that are connected or perhaps confused with it. For instance, if you were defining prions, you might want to distinguish them from viruses; if you were defining a modern version of evolutionary theory, you should contrast it with classic Darwinian theory.

CLASSIFYING

Classifying means dividing something into its separate parts according to some principle of selection. This principle or criterion may vary. You could classify crops, for example, according to how they grow (above the ground or below the ground), how long they take to mature, or what climatic conditions they require. The population of an ecological niche might be classified according to species present, distribution within the niche, degree of interdependence, and so on.

If you are organizing your essay by a system of classification, remember the following guidelines:

- All members of a class must be accounted for. If any are left over, you need to alter some categories or add more.
- Categories can be divided into subcategories. You should consider using subcategories if there are significant differences within a category. For instance, if you were classifying the effects of brain damage according to the site of the lesion, you might want to create subcategories for memory deficits, language problems, perceptual difficulties, and so on.
- Any subcategory should contain at least two items.

EXPLAINING A PROCESS

This kind of organization shows how something works or has worked, whether it be the weather cycle, the photochemical reactions of the eye, or the mechanisms of hibernation. The important point to remember is to take a systematic approach to break down the process into a series of steps or stages. Although your order will vary depending on circumstances, most often it will be chronological, in which case you should see that the sequence is accurate and easy to follow. Whatever the arrangement, you can make the process easier to follow if you start a new paragraph for each new stage.

TRACING CAUSES AND EFFECTS

Tracing causes and effects is essentially the same as explaining a process but with an emphasis on showing how certain events have led to or resulted from other events. Usually you are explaining why something happened. Showing how certain events have led to or resulted from others is a complex process, and you must be careful not to oversimplify a cause-and-effect relationship. If you are tracing causes, distinguish between a direct cause and a contributing cause, between what is a condition of something happening and what is merely a correlation or coincidence. For example, if you discover that there is a high correlation between the number of garbage dumps and the number of polar bears sighted around Churchill, Manitoba, you should not conclude that the bears breed in the garbage dumps. Similarly, you must be sure that the result you mention is a genuine product of the event or action.

COMPARING

Many successful essays are based on comparisons. Even if it is not specifically part of your assignment, by choosing a limited number of items to compare—

theories, neural mechanisms, or factors that might influence behaviour, for example—you can create a clear focus for your paper. Just be sure that you deal with differences as well as similarities.

When you prepare your outline for this kind of paper, you will have to decide how best to set up your comparisons. The easiest way of comparing two things—though not always the best—is to discuss the first subject in the comparison thoroughly and then move on to the second. The problem with this kind of comparison is that it often sounds like two separate essays slapped together.

A more effective approach is to integrate the two subjects, first in your introduction (by placing both in a single context) and again in your conclusion, where you should bring together the important points you have made about each. When discussing the second subject, you should always refer back to your discussion of your first subject ("Unlike butterflies, moths are nocturnal and hold their wings outstretched when at rest. . . ."). This method may be the wisest choice if the subjects you are comparing seem so unlike that it is hard to create common categories in which to place them for discussion; that is, if the points you are making about *X* are of a different type from the points you are making about *Y*.

If you can find similar criteria or categories for discussing both subjects, however, the comparison will be most effective if you organize it so that you make all of your comparisons within one category before moving on to the next. Because this kind of comparison is more tightly integrated, it is easier for the reader to see the similarities and differences between the subjects. As a result, the essay is likely to be more forceful.

INTRODUCTIONS

The beginning of a research paper has a dual purpose: to indicate your topic and the way you intend to approach it, and to whet your reader's appetite for what you have to say. One effective way of introducing a topic is to place it in a context—to supply a kind of backdrop that will put it in perspective. The idea is to step back a pace and discuss the area into which your topic fits, and then gradually lead into your specific field of discussion. Sheridan Baker (Baker & Gamache, 1988) calls this the *funnel approach* (see page 69). For example, suppose that your topic is the specific action of certain neurotoxins on the nervous system. You might begin with a general discussion of the effects of some of the naturally occurring neurotoxins, such as the one produced by the puffer fish (also known as the fugu fish, which is served in certain Japanese restaurants). You would then narrow the focus of your discussion to the point where you can move into the specific topic you plan to discuss, the effects of neurotoxins on the nervous system. The funnel opening is applicable to almost any kind of paper.

You should aim to catch your reader's interest right from the start. You know from your own reading how a dull beginning can put you off a book or an article. The fact that your instructor has to read on anyway makes no difference. If a reader has to get through thirty or forty similar essays, it's all the more important for yours to stand out. However, it's also important that your lead-in relate to your topic: never sacrifice relevance for catchiness. Finally, whether your introduction is one paragraph or several, make sure that by the end of it your reader knows exactly what the purpose of your essay is and how you intend to accomplish it.

CONCLUSIONS

Endings can be painful—sometimes for the reader as much as for the writer. Too often, the feeling that one ought to say something profound and memorable produces either a pompous or a meaningless ending. You know the sort of thing:

> This is not to say that the research of the last decade has been in vain, to the contrary, we have learned a tremendous amount. Critical and conflicting evidence does tell us however, that we still have a long way to go before we can deal in absolutes.

Even if you ignore the grammatical difficulties in this example, it easy to see that these two sentences could easily have been omitted without affecting the substance of the paper. Experienced editors say that many articles and essays would be better without their final paragraphs: in other words, when you have finished saying what you want to say, sometimes the best thing to do is stop. This advice may work for short papers, where you need to keep the central point firmly in the foreground and don't need to remind the reader of it. However, for longer pieces, where you have developed a number of ideas or a complex line of argument, you should provide a sense of closure. Readers welcome an ending that helps tie the ideas together so that they don't feel as though they've been left dangling. And since the final impression is often the most lasting, it's in your interest to finish strongly. Simply restating your thesis or summarizing what you have already said isn't forceful enough; the following are some of the alternatives:

THE INVERSE FUNNEL

The simplest conclusion is one that restates the thesis *in different words* and then discusses its implications. Baker & Gamache (1988) call this the *inverse funnel*

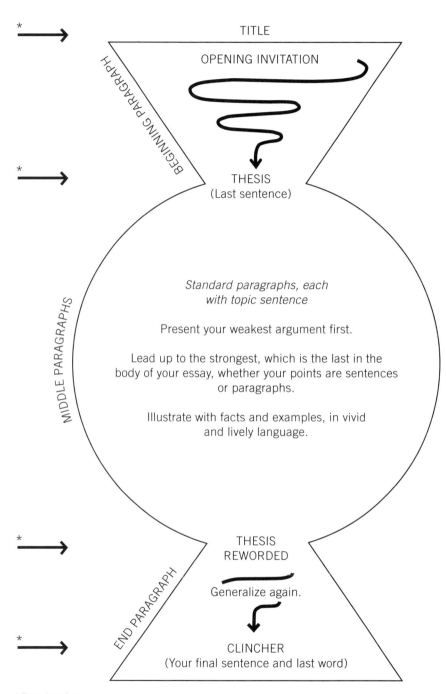

*Focal points

SOURCE: Sheridan Baker and Laurence B. Gamache (1988, p. 65). Reprinted by permission of Pearson Education Canada Inc.

approach, as opposed to the funnel approach of the opening paragraph. The danger in moving to a wider perspective is that you may try to embrace too much. When a conclusion expands too far it tends to lose focus, as in the example above. It's always better to discuss specific implications than to trail off with vague generalities. So, in the conclusion to your paper on neurotoxins, you would begin with a summary of the effects of neurotoxins on the nervous system, followed by a brief comment on the advisability of eating fugu fish as a delicacy.

THE FULL CIRCLE

If you began your essay by relating an anecdote, citing some unusual fact, or raising a rhetorical question, then you can complete the circle by referring to it again in your conclusion. At this point, you can relate it to some of the insights revealed in the body of the essay. This technique provides a nice sense of closure for the reader.

THE STYLISTIC FLOURISH

Some of the most successful conclusions end on a strong stylistic note. Try varying your sentence structure to make the conclusion stand out from the rest of the text. So, if most of your sentences have been long and complex, make the last few short and emphatic. Be careful not to overdo it, though. If you're not comfortable modifying your style, you might be better off playing it safe and sticking with one of the methods that's more conducive to your writing style.

THE EDITING STAGE

Often the best writer in a class is not the one who can dash off a fluent first draft, but the one who is the best editor. To edit your work well you need to see it as the reader will, and in order to do that you have to distinguish between what you meant to say and what you have actually put on the page. For this reason it's a good idea to leave some time between drafts so that when you begin to edit you will be looking at the writing afresh, rather than reviewing it from memory. This is the time to go to a movie or do something that will take your mind off your work; without this distancing period you can become so involved in your paper that it's hard to see it objectively.

Editing doesn't mean simply checking your work for errors in grammar or spelling. It means looking at the piece as a whole to see if the ideas are *well organized*, *well documented*, and *well expressed*. It may mean making changes to the structure of your essay by adding paragraphs or sentences, deleting others, and moving others around. Experienced writers may be able to check several aspects of their work at the same time, but if you are inexperienced or in doubt about your writing, it's best to look at the organization of the ideas before you tackle sentence structure, diction, style, and documentation. As you read through your paper, ask yourself whether it flows properly: is the rhythm comfortable, or do some passages slow you down because they're awkwardly constructed? Such problems can often be corrected just by shuffling the word order a little.

It's also a good idea to try to take on the role of your reader. Keep asking yourself whether your explanations are sufficiently clear. Would they make sense to a reader who is less familiar with the material than you? Are there places where you have assumed the reader would know what you're referring to when you make a statement? Remember that you should think of your reader as someone who is intelligent and educated, but who does not necessarily share your level of expertise in the subject you are discussing.

What follows is a checklist of questions to ask yourself as you begin editing. This is not an all-inclusive list but focuses only on the first step: examining the organization of your work. Since you probably won't want to check through your work separately for each question, you can group some together and overlook others, depending on your own strengths and weaknesses as a writer.

AN EDITING CHECKLIST

- Is my title concise and informative?
- Are the purpose and approach of this essay evident from the beginning?
- Are all sections of the paper relevant to the topic?
- Is the organization logical?
- Will the subheadings be meaningful to the reader? Do they clearly identify the various sections of the paper?
- Do the paragraph divisions give coherence to my ideas? Have I used them to keep similar ideas together and signal movement from one idea to another?
- Do any parts of the paper seem disjointed? Should I add more transitional words or logical indicators to make the sequence of ideas easier to follow?
- Do my transitional words really link two ideas with a logical connection or are they just filling in space?
- Are the ideas sufficiently developed? Is there enough evidence, explanation, and illustration?

- Would an educated person who hasn't read the primary material understand everything I'm saying? Should I clarify some parts or add any explanatory material?
- In presenting my argument, do I take into account opposing arguments or evidence?
- Have I been accurate and fair in my representation of what my sources say?
- Have I cited all the sources I used? Is the style of in-text citations consistent and appropriate?
- Do my illustrations and figures add anything to the paper? Do they present information in the clearest, most effective way? Have I described or discussed each one clearly and completely?
- Have I checked the illustrations and figures for spelling and content? Have I cited the sources?
- Are the tables complete and correct, with the sources noted?
- Do my conclusions accurately reflect my argument in the body of the work?
- Is my reference list accurate and complete? Is it in the correct format for my discipline?

Another approach would be to devise your own checklist based on comments you have received on previous assignments. This is particularly useful when you move from the overview of your paper to the close focus on sentence structure, diction, punctuation, spelling, and style. If you have a particular weak area—for example, irrelevant evidence, faulty logic, or run-on sentences—you should give it special attention. Keeping a personal checklist will save you from repeating the same old mistakes.

REFERENCES

Baker, S. & Gamache, L.B. (1988). *The Canadian practical stylist.* (4th ed.). Don Mills, ON: Addison Wesley.

McCrimmon, J. (1976). *Writing with a purpose* (6th ed.). Boston: Houghton Mifflin.

Trimble, J.R. (1975). *Writing with style: Conversations on the art of writing.* Englewood Cliffs, NJ: Prentice-Hall.

cHApter 7

Writing a Lab Report

ORGANIZATION AND FORMAT

If you are studying psychology or biological sciences, much of the writing you do will take the form of *lab reports*—that is, formal descriptions of experiments you have done. Lab exercises have several purposes, but the most important one is to give you a "hands-on" introduction to experimentation.

In most lab courses you will receive a manual outlining what you will be doing in the course. You should always look through it to see how the course is structured and what will be coming up later. Knowing what lies ahead may help you to organize your early assignments. An especially important bit of preparation you should do is reading the required material before you go to the lab. Although the lab itself may not be difficult, there are always lots of details for you to grasp, especially if you are not familiar with routine lab procedures. In psychology, this will almost certainly be the case, and, because you are likely to be dealing with human participants, you will need to be extra careful to ensure things run smoothly. You will also need to keep track of all of the experimental details in preparation for your write-up.

One of the more important lessons you will learn from a lab course will come when you begin to write your lab report. When you read a description of a study in a journal, especially if it is well written, you are presented with an eloquent account of what appears to have been a flawless procedure. In reality, experiments rarely go exactly according to plan. That means that a researcher may begin by doing one thing and end up doing something quite different. However, these complications with and departures from the original plan are seldom apparent in the published paper. As a student in psychology or the life sciences, you will have to learn how to distinguish those aspects of your lab work that are relevant and should be included in a write-up from the snags and false starts that are not informative to the reader and should be omitted.

Writing a lab report is a little different from writing an essay in the Arts and Humanities. Scientists are interested in the orderly presentation of factual evidence to support hypotheses or theories. This means that the structure of reports in the sciences will tend to be more formal than in non-science disciplines. It also means that you must be objective in the way you report your data. Although you may wish to make a case for a particular hypothesis, it is essential to separate the data themselves from your own speculations about them. You must present your information so that anyone who reads your report can understand exactly what you've done in your experiment. On the basis of the evidence you present, the reader should be able to draw his or her own conclusions; if you've done a good job, they will be the same as yours.

PURPOSE AND READER

Whether you are writing for a prestigious journal or for a graduate-student lab instructor, your goal is to disseminate scientific information. As an undergraduate, you will be writing lab reports to describe studies you have conducted and to demonstrate that you understand a particular phenomenon or theory. You can assume that the reader—your instructor or, later on in your career, a colleague or peer—is familiar with basic scientific terms, so you won't need to define or explain them. However, you cannot assume that your reader is all-knowing. He or she will be frustrated to be told that "The participant's settings were read directly from the micrometer drive" if this is the first time a micrometer has been mentioned anywhere in the paper. Be sure, then, that you have introduced procedures or pieces of equipment before you start making references to them. You can also assume that your reader will be on the lookout for any weaknesses in methodology or analysis, and any omissions of important data.

OBJECTIVITY

Objectivity is essential in lab work. You should never let your preconceived opinions or expectations interfere with the way you collect or represent your data. If you do, you run the risk of distorting your results, perhaps even unwittingly. We have already talked about the ethics of manipulating data, but you need to be aware that if there is some ambiguity about a piece of data, your decision to accept or reject it may be influenced by the way you think the results should turn out. You should conduct your experiment as objectively as possible and present the results in such a way that anyone who reads your lab report would be likely to reach the same conclusions that you did. Although we always think that we can be objective, our wishes and ambitions can sometimes get in the way. Allegra Goodman (2006), in her novel *Intuition*, gives a fascinating and quite credible account of how pressure to be successful in the battle to gain recognition and grant funding can have serious effects of the judgement of otherwise honest individuals. As we mentioned earlier, outright fraud is probably rare, but narrow thinking and unwillingness to accept evidence that does not conform to your thinking is not.

THE STRUCTURE OF A LAB REPORT

Although the general format of lab reports is similar for all disciplines, formal style requirements vary considerably. Your lab instructor will give you the specific information for your discipline. In addition, however, you can obtain a lot of information about both style and format from the *Publication Manual of the American Psychological Association* (American Psychological Association, 2001) and *Scientific Style and Format: The CSE Manual for Authors, Editors, and Publishers*, published by the Council of Science Editors 2006. The *CSE Manual* is most useful for those working in disciplines where it is important to be accurate when describing different species of plants or animals or different chemical compounds. Most of the suggestions that follow are adapted from the *APA Manual*, and we refer you there for more details.

Because the information in scientific reports must be easy for the reader to find, it should be organized into separate sections, each with a heading. By convention, most lab reports follow a standard order:

1. Title page
2. *Abstract*
3. *Introduction*
4. *Method* – this may include some or all of the following subsections:
   • *Subjects* (or *Participants*)
   • *Apparatus* (or *Materials*)
   • *Design*
   • *Procedure*
5. *Results* (including figures and tables)
6. *Discussion*
7. *References*
8. *Footnotes*

The order of these sections is always the same, although some sections may be combined or given slightly different names, depending on how much information you have in each one. Different disciplines also have slightly different rules, but the following will give you an overview of what should go into each section of your report.

TITLE PAGE

The first page of your report is always the title page. It should include the title of the paper, your name and student number, the name and number of the course, the instructor's name, and the date of submission. Your title should be brief—no more than 10 or 12 words—but informative, and it should clearly describe the topic and scope of your study. Use words in the title that you might use as keywords if you were doing a literature search for studies on your topic. Avoid meaningless phrases, such as "A Study of . . ." or "Observations on . . .". Simply state what it is that you are studying: for example, *Effects of Gamma Rays on Growth Rate of Man-in-the-Moon Marigolds*. Sometimes you may want to emphasize the result you obtained: for example, *Brief Exposure to Gamma Rays Increases Growth Rate of Man-in-the-Moon Marigolds*.

ABSTRACT

The *Abstract* appears on a separate page following the title page. This is a brief but comprehensive summary of your report that should be able to stand alone; that is, someone should be able to read it and know exactly what the experiment was about, what the results were, and how you interpreted them. For a professional researcher, the *Abstract* is arguably the most important section of a paper because it is the first point of access in a literature search. If the *Abstract* does not attract the reader's interest, the whole report is likely to be ignored. For this reason, the *Abstract* should include all the major points of your study—and exclude anything that is not in the report itself. In no more than roughly 120 words it should describe the purpose of the experiment, the participants, the experimental apparatus or materials, the procedure, the main results (including statistical significance levels), and your conclusions. Because the *Abstract* is so short, you don't have space to be vague. Avoid wordy phrases—such as, "The reason for conducting the experiments in this study of X was to examine . . ."—when you can be more concise: "I studied X to examine . . .". At the end, rather than state the obvious—for example, "The study produced some very interesting results from which we can draw several conclusions . . ."—state directly what your conclusions are: "The results show . . ." or "I conclude that . . .".

INTRODUCTION

The *Introduction* describes the problem you are studying, the reasons for studying it, and the research strategy you will use to obtain the relevant data. This is also where you will present your experimental hypothesis and a statement of either what you expect to find or the question you plan to examine. Not all introductions contain explicit predictions; some simply present a question the researcher hopes to answer. For example, if you were interested in the social development of infants, you might ask about the age of a baby's first smile and how the frequency of social smiling changes with age. If you don't make a specific prediction, you should state your question clearly so that your reader knows exactly what the purpose of your study is.

You should also explain the background to your topic in the form of a brief review of the relevant literature. This should not be an exhaustive discussion but rather an overview that recognizes the prior work of others and shows how your own study relates to what has come before. You don't need to summarize all the aspects of the studies you cite, only those points that are relevant to your own study, including, if appropriate, the theory underlying your experiment. If, as is often the case, the purpose of your study is to test a hypothesis about a specific problem, you should state clearly both the nature of the problem and what you expect to find.

The final paragraph or so of your *Introduction* should be a summary of what you did in your experiment. It should include a description of the variables that you manipulated, a formal statement of your experimental hypothesis, and a brief explanation of the reasons why you expected to get a particular pattern of results.

METHOD

The *Method* section is usually made up of several labelled subsections. These describe the organisms you were working with, your experimental apparatus and materials, and your procedure.

SUBJECTS (PARTICIPANTS)

In any discipline that studies live organisms, you need a section describing your experimental animals. This section is usually called *Subjects* unless these animals happen to be human, in which case it is more likely to be called *Participants*.

If human participants are involved in your study, you need to give any information about them that is relevant to the experiment. Including this information in your report is important because the extent to which you can generalize your results depends on how representative your sample is of a population. For example, if you were studying people's judgments of the meanings of certain kinds of words, it would be important to mention whether or not all the participants were native speakers of English. Typically, you would also give the status of each of the participants and the average age of the group ("university undergraduates" or "3-year-old children registered in a pre-school program").

Whenever you're in doubt about whether to include certain information, ask yourself: "Is it relevant to the purpose of the study?" This is a good way to make sure you stick to the essential details.

When your experiment involves non-human subjects, you should mention the species or strain, the sex, if appropriate, and any special characteristics they might possess. For example, there are specific strains of rat that have been bred to have a strong preference for alcohol. If you were conducting an alcohol study, you would need to indicate what kind of rats you were using.

APPARATUS (MATERIALS)

Depending on the discipline and the kind of experiment you are doing, this section may be entitled either *Apparatus* or *Materials*; consult with your instructor to find out the rules that apply to your own experiment.

This section should contain a description of the materials and any experimental apparatus you used. You should describe the essential components of any major pieces of equipment you used and how they were set up. If you used different arrangements of the equipment for different parts of the experiment, give a full list of the equipment in this section, and in the *Procedure* section, describe each separate arrangement before you describe the procedure it was used for.

If the equipment is a standard, commercially available item, it is customary to give the manufacturer's name and the model number. For example, if you were required to display patterns on the face of an oscilloscope cathode ray tube for a psychology experiment, you should say that you displayed the patterns "on a Tektronix 608 display monitor with a P31 phosphor," or whatever it was that you used.

Nowadays, most experiments involve some use of computers, so you should indicate what kind of computer was used in your experiment (e.g., Macintosh G4). If your computer needs were more specialized—for instance, if you used the computer to generate visual stimuli or if you had a special kind of sound card to generate a particular combination of sounds—then you must provide additional information about these components, such as the name and model number of the monitor or the specific kind of sound card.

DESIGN

Sometimes, especially if your study involves a fairly complex experimental design, it is helpful to include a brief, formal description of that design. In general terms, "design" refers to the way in which you will run your experiment— what experimental conditions you will use, how many participants you will include in each group, and so on. But "design" also refers to the statistical model you will apply to your experiment. In this section of your lab report, you should describe that model in relation to your experiment. For instance, let's say you wanted to run an experiment to study the effects of room temperature on the ability to perform three different tasks. You might choose to run two groups of participants, one working in a high temperature, the other in a low temperature; each group would perform the same three tasks. Such an experiment is referred to as a *between-within* design, because when you analyze your experiment, you would compare overall difference in performance *between* the two groups and you would look at whether performance on individual tasks differed *within* each temperature condition. These data would then be analyzed using a particular class of Analysis of Variance. You would need to select your experimental conditions to fit the requirements of this design and describe these in your *Design* section.

The *Design* section should include descriptions of

- your independent and dependent variables, indicating which are between-subject variables and which are within-subject variables;
- the composition of your experimental and control groups, and how the subjects were assigned to the groups; and
- the statistical model: for example, a two-way factorial, or a repeated measures design.

PROCEDURE

This section is a step-by-step description of how you carried out the experiment. If your experiment consisted of a number of tests, you should begin this section with a short summary statement listing the tests so that the reader will be prepared for the series. When you describe the tests later on in this section, discuss them in the same order you used to list them, to avoid confusion.

The *Procedure* section must be written with enough detail that others would have no difficulty repeating the experiment in all its essential details. However, you should avoid any information that is not directly relevant to the study. If you were measuring the effectiveness of different antibiotics on bacterial growth, you would have to describe the medium on which you grew the bacteria, the incubation temperature, the strength and amount of antibiotic given, and so on; you would not need to mention the fact that the experiment was carried out on the fifth floor of the Biology building.

As a general rule, if you are following instructions in a lab manual, you should not copy them out word for word because this might be considered plagiarism. If you want to be absolutely clear on this issue, ask your lab instructor what the policy is about using wording from your lab manual in your report.

When describing experiments, it is standard practice to use the past tense. However, there has been some debate among scientists who write for scientific journals about the use of active or passive voice (e.g., "*I placed* the rat in the water maze" versus "*the rat was placed* in the water maze"). Traditionally, only the passive voice was used for this kind of writing because its more formal, detached quality was considered appropriate for a scientific report. More recently, there has been a tendency to emphasize the active voice because it is clearer and less likely to produce awkward or convoluted sentences. The *Publication Manual of the American Psychological Association* (2001) advocates the use of the active voice; however, you should also consider what sounds best in the context of the paper. A sentence like "I placed the rat in the water maze" sounds a little bit too self-conscious, so the passive may be more appropriate here. On the other hand, it is more awkward to say, "Questionnaires were required to be filled out by the participants . . ." than to say, "Participants had to fill out questionnaires . . .". Use your judgment about what sounds best to your ears.

RESULTS

When professional scientists come across a new paper, they will read the *Abstract* first. After that they may only glance at the *Introduction* and the *Method* section before focusing their attention on the *Results* section, as this is where they will find the essence of the paper. In a sense, all the other sections of a lab report are subordinate to the *Results*; it is here that readers will find the most important information. They may have questions about the methodology, and they may disagree with the interpretation presented in the *Discussion*, but they should be able to make their own evaluation of the findings on the basis of a *Results* section that presents the data clearly and unambiguously. For this reason, you should spend some time thinking about the best way to present your results.

The *Results* section should contain a summary of all the data you collected, with sufficient detail to justify your conclusions. You might include some results in tabular form and others in figures. Typically, you will provide statistical summaries of your results rather than data on individual subjects. It is important that you present all of your data, whether or not it supports your hypothesis. Although the *Results* section is not the place to discuss the implications of your data, it is appropriate to guide your reader through the findings. So, for example, you might say:

> The first question was whether listening to loud music on a personal stereo impaired performance in a word-search task. The average time

taken to complete the task in the respective experimental conditions was _____ [*here you would give a description of your data and statistical analyses*]. These results are consistent with the experimental hypothesis.

Without discussing why the results turned out the way they did, you have provided the reader with a context in which to place the data you describe. This is much easier to read than a dry listing of means and statistical tests without any explanation.

Because the *Results* section is so important, Chapter 9 of this book is devoted to the more technical aspects of presenting your data, including the preparation of tables and figures.

DISCUSSION

This part of the lab report allows you the greatest freedom because it is here that you will examine and interpret your results and comment on their significance. You will want to show how the experiment produced its outcome—whether expected or unexpected—and to discuss those elements that influenced the results.

Before beginning your actual discussion, you should give a brief overview of the major findings of your study: for example, "The results of the present study demonstrated that university students could remember the details of lectures much better if they spent fifteen minutes organizing and expanding their lecture notes at the end of each class." The rest of the discussion would deal with the reasons why this might be so.

To help you decide what to include in the *Discussion* section, you might try to answer the following questions:

- Do the results reflect the objectives of the experiment?
- Do these results agree with previous findings, as reported in the literature on the subject? If not, how can you account for the discrepancy between your own data and those of other students and scientists?
- What, if anything, may have gone wrong during your experiment, and why? What was the source of any error?
- Could the results have another explanation?
- Did the procedures you used help you to accomplish the purpose of the experiment? Does your experience in this experiment suggest a better way for next time?

The order of topics in your *Discussion* section should be the same as that in your *Results* section. Discuss each of your findings in turn as you reported them in that section. If you have a result that you can't explain, say so; never ignore an inconvenient finding in the hope that your instructor might not notice. A good *Discussion* section may not be able to tie up all the loose ends, but it must acknowledge that these loose ends exist.

You should always end with a statement of the conclusions that may be drawn from the experiment. Sometimes the conclusions are put in a separate section, but typically they form the final paragraph of the discussion. You might end your discussion on the fictitious study above by saying: "The findings of the present study suggest that if students took a little extra time going over their notes at the end of each class, it is likely that they would improve their grades."

Sometimes, especially if the discussion is going to be straightforward, you can combine the *Results* and *Discussion* sections. In this case, the best strategy is to present each result, followed by a brief discussion. At the end, you should try to pull it all together in a concluding paragraph.

REFERENCES

We have already discussed plagiarism in Chapter 4. The way to avoid any suspicion of plagiarism is to support every non-original statement with a reference citation. Always refer to your sources, unless the information you are providing is considered common knowledge. Each time you refer to a book or an article in the text of your report, cite the reference; then at the end of the paper make a list of all the sources you have cited. The precise form of the citations and reference list varies among disciplines; for more detail, see Chapter 11.

FOOTNOTES

You should use footnotes as little as possible. If they are unavoidable, indicate where each one should go in the text by placing a superscript reference number at the point of insertion. At the back of the paper, after the references, list the footnotes in order, numbering them so that they correspond to the numbers you have used in the text.

REFERENCES

American Psychological Association. (2001). *Publication manual of the American Psychological Association* (5th ed.). Washington, DC: American Psychological Association.

Council of Science Editors. (2006). *Scientific style and format: the CBE manual for authors, editors, and publishers*. Bethesda, MD: Council of Science Editors.

Goodman. A. (2006). *Intuition*. New York: The Dial Press.

chApter 8

RESEARCH PROPOSALS AND THESES

One reason for requiring you to write lab reports in your early years at university is to prepare you for a much more extensive project that you may do in your senior years. At some institutions you may be expected to do an honours thesis involving an independent project; elsewhere you may be required to prepare a formal research proposal without actually carrying out the study itself. Although you can apply some of the guidelines we've provided for writing lab reports and research papers to preparing research proposals and theses, projects of this kind are sufficiently different to warrant their own chapter. They will require more planning and independent thought on your part, and they will have a somewhat different organization from a simple lab report or essay.

THE RESEARCH PROPOSAL

Although it is not likely that you will have to prepare a formal research proposal for your undergraduate thesis, it is often a requirement once you are in a graduate program. If you go on to be a professional scientist, then the research proposal will be in the form of an application for a grant. Because grant funds are the lifeblood of any research scientist, the application is a crucial document. The earlier you learn the skills involved in applying for grants, the more likely it is that your application will be successful. Learning to prepare a research proposal is a first step.

Writing a research proposal serves a number of purposes:

- It allows you to do a focused literature search within a specific research area.
- It gives you an opportunity to think through all aspects of a research problem before you begin.
- It gives you an opportunity to convince your reader that the project is worthwhile.

You can think of the proposal as a strategic plan for a research project. As such, it should have three elements:

1. a statement of the problem you intend to study;
2. a statement of why it is important to study it; and
3. a statement about how you will carry out your study.

The research proposal combines elements of both an essay and a lab report. The introductory or background section presents a critical review of a body of literature with the addition of a statement of your experimental hypotheses. The rest may resemble a lab report written in the future tense.

As with a lab report or an essay, your proposal will have a number of parts. We describe the essential components of a proposal below, but you may prefer to rearrange the order of these parts for the proposal you write.

INTRODUCTION

Your introduction should include sections that consider the following:

- **Context of the research project.** The first thing you need to do is to set the stage. Describe briefly the current thinking in the area you will be working in.
- **Review of the literature.** You need to establish the relationship between your proposed work and the existing literature in the field. Here, you are moving from the current thinking in the general area you will be working in to the opinions set forth in the specialized literature that are directly relevant to your own project. This review does not have to be exhaustive; rather, it is intended to show the reader that you are familiar with the research literature in the area.
- **Statement of the problem.** Once you have established a context, then you need to introduce your own study. Indicate what it is that you plan to study and what the rationale is. If you have any underlying assumptions, you should state them here. Finally, you should state what your experimental hypotheses are, or what questions you hope to answer.

METHODOLOGY

This part of a research proposal should include sections on research design and data analysis. In this section you should indicate what kinds of data you will collect, the methods you will use to gather the data, and the rationale for using these particular methods. The amount of detail you provide here will depend on the space limitations of the proposal, but you should always include enough information to make the reader aware that you know what you are doing.

You should also indicate the formal experimental design you will be using. You can avoid a lot of problems if you know exactly how you will analyze your data *before* you run your study. So, if each participant will be tested under both experimental and control conditions you should be using a within-subjects design. Other experiments might require a multifactorial analysis of variance design or multiple correlations. Although it is not essential for your proposal, you should also be thinking of how you will collate and organize your data as you collect it so that it will be a simple procedure to carry out the analysis.

ANTICIPATED RESULTS
You can think of this section as the discussion section of a lab report written in the future tense. For the proposal, you do not need to place the results in the context of the literature. Instead, you should draw out the implications or the significance of your results if they turn out according to your predictions.

LIMITATIONS
Few studies are perfect. If you are carrying out a project as an undergraduate thesis, then it's likely that you will not have all the resources that a fully funded research professor might have. As a result, you may not be able to carry out your study exactly as you would like, and the results you obtain might not be perfect. Such flaws and limitations will not be held against you, but you must indicate to your reader that you are aware of them. You might also have to concede that there may be alternative explanations for your study and the resulting data. Acknowledging the potential problems with your study will not weaken your proposal, as long as you can defend everything you plan to do.

If you have aspirations to be a professional scientist, then learning how to write a proper research proposal may be one of the most important skills you acquire as an undergraduate. If you are applying for a grant as a new investigator, the adjudication committees have little to go on other than what you have written in your proposal. If you can develop a convincing style and combine it with logical thinking, you will be well on your way to being a successful, and well-funded, scientist.

HONOURS THESES

In a typical lab course you will be assigned several experiments to carry out and write up. Occasionally you may get a chance to do an independent project, but most of the time you will be working under close supervision. As you progress through your university career, however, you will be required to do more independent work; this will sometimes include writing an honours thesis. In their final format, lab reports and theses are quite similar, and many of the rules we described in Chapter 7 apply also to theses. However, the thesis is a much bigger project, and it requires a good deal more preparation.

Although the primary purpose of this book is to help you express yourself well in writing, it's important to keep in mind that what you finally write is largely a product of what you have done before you get to the writing stage. In the case of an honours thesis, you will be spending a whole academic year preparing to write a single paper, so it is all the more important that you think carefully about how you'll proceed. It's surprising, but some students do not have a clear idea of what is required for an honours thesis until they have actually begun work on it, and in some cases, not until they have been working on it for quite a while. This section will give you a brief overview of what is required to prepare for an honours thesis involving an empirical research project. Knowing what is expected of you from the beginning will make it much easier to write a final paper that will earn you a good mark.

WHAT IS AN HONOURS THESIS?

An honours thesis is an opportunity for you to demonstrate that you are capable of carrying out independent scientific research (albeit with some guidance from your adviser) and presenting it in the form of a paper. It is equally important, however, to be aware of what a thesis is not.

An honours thesis is not primarily intended to be an original and significant contribution to the literature of your discipline. Although many theses are excellent pieces of work, and some of the best are good enough to get published, this is not the reason for doing the work. You must not assume that you have to create a masterpiece that solves a "big" problem in your discipline. Many students get bogged down after designing complicated experiments involving many dozens of subjects; as a result, they run out of time and cannot do a good job on their data analysis or write-up. Research is time-consuming, and it often involves many false starts. When you are doing a thesis over the course of a single year, you don't have the flexibility to go back and start again if things don't work out. For this reason it's essential to set reasonable goals and choose a project that will be manageable. You can demonstrate your research skills just as well—sometimes better—with a simple project as with a complicated one.

You should also be aware that the specific topic area in which you choose to do your thesis is less important than you might think. For example, many psychology students who aspire to go on to graduate programs in clinical psychology naturally want to do their theses in this area. However, clinical psychology is, in fact, one of the most difficult areas in which to do meaningful research in a short period of time. Several obstacles can slow down this kind of research: it may be difficult to get ethics clearance; the potential subject population may not be readily available; and the rules of confidentiality may prevent a student from gathering important data. It is possible (with the right adviser and a carefully chosen project) to produce an excellent thesis in an area such as clinical psychology, but you may have to work a lot harder than, say, a student doing a cognitive psychology thesis. Similarly, students who do animal research often find that although the project may be straightforward, they have to commit themselves to many hours of testing, often over weekends and holidays.

The important thing is to choose a feasible project that you will find interesting. If you keep this in mind, the task of selecting a topic and running the study will be much easier.

DIFFERENCES BETWEEN A LAB REPORT AND A THESIS

The main differences between a lab report and a thesis are the amount of background that you will be expected to provide in your *Introduction*, the scale of your study, and the extent of your *Discussion* section. Otherwise, the final write-up of a thesis should follow a format similar to the one used for a lab report. One advantage in writing the thesis is that you will often have the chance to submit preliminary drafts to your adviser for comments before you write the final version. Remember, though, that to get feedback you have to submit the drafts in time for someone to read them. Don't leave everything until the last minute!

Depending on your adviser, you may be assigned a specific topic or given a general topic area in which to work, or you may be told to start reading the literature and come up with your own topic independently. The more freedom you are given, the more important it becomes to organize your time. Think about all the steps you have to take, think about your own strengths and limitations, and plan accordingly. For instance, if you know that statistics is not your forte, plan to spend more time on data analysis. Whatever you do, do not procrastinate. No matter how well organized you are, it is almost certain you will find yourself rushed at the end. Don't let a whole year's work get graded down just because you didn't give yourself enough time to write a polished final draft.

The best way to start is to make up a schedule outlining when different aspects of your project should be completed. Consult with your adviser, because he or she will know which parts of your project are most likely to slow you down. The specifics of the schedule you come up with will depend on the project you are doing, so we will not try to advise on that. Instead, we have listed below the major stages you will have to complete.

Background reading

Even if your adviser has assigned you a topic, you will need to do some background reading before you are ready to prepare a research proposal. Your adviser will probably give you some papers to start with, but then you will have to do a proper literature search and then spend time going through that material.

Formulating a project and preparing a research proposal

Although your adviser may not ask for a formal proposal, it is to your advantage to have one. This is something that you may be able to use later on as part of your *Introduction*, and it will also serve as a guide as you proceed. You can follow the suggestions made earlier in this chapter about preparing research proposals.

Getting ethics approval

If you are running a project that involves testing animals or humans, you will need to have it approved by an ethics committee to ensure that it is considered ethical. Different institutions have different rules in this regard, so check with your adviser.

Setting up the experiment

Your proposal should contain the formal experimental design for your study. This stage involves working out what materials and apparatus you will need to run the experiment. For example, do you need to have experimental apparatus constructed? Is there new software that you need to become familiar with? Are the tests or questionnaires you plan to administer readily available? There will also be other questions about your experiment to consider at this stage. For instance, how will you recruit the participants for the study? How should you arrange your data sheets for ease of entry? How long will an experimental session last? These are only some of the issues you will need to consider before you begin your experiment. What you must do, though, is work through them all so that you will be prepared once the study begins.

Defining the experimental protocol

The final step before you begin to run your study is to work out exactly the procedure you will follow, starting from the moment the participant walks into the lab. It is worth writing down, in detail, precisely what it is that you will be doing, for two reasons. First, this preparation will give you confidence once you begin to run your study. Second, when it comes time to do your final write-up, you will have your own detailed account to work from.

RUNNING THE EXPERIMENT

In some ways, running the experiment is the scariest part of the study. This is where you want everything to run smoothly. If you have prepared yourself, rehearsed the way that you will do your testing, and, ideally, tested a couple of pilot subjects and made any necessary changes to the experimental protocol, you will know what to expect and shouldn't encounter any unforeseen problems.

DATA ANALYSIS

Once you have collected your data, you will need to analyze it. At one time, this would have been a complicated and time-consuming procedure. Nowadays, as long as you have planned your experiment carefully, it should be very straightforward: all you need to do is plug your numbers into an analysis program and see how it turns out. Be aware, though, that even though it is now very easy to use commercially available software to analyze your data, it is your responsibility to know what these programs are doing so that you can explain the analysis to a reader or to the members of an oral examination committee. Your findings will not be as convincing if you cannot explain how the software processed your data to generate the results it did.

For some studies, you can summarize your data in the form of tables. For others, you may need to draw graphs. It is a very useful exercise to prepare draft copies of your tables and graphs based on the kind of data you expect to get. Doing this will help you think through your experiment before you run it. It will also allow you to consider exactly how to fine-tune your experimental conditions to show off the results in the best possible light.

WRITING UP THE THESIS

Logically, writing the thesis is the final stage of your research project. However, if you are well organized, you will have at least drafted a significant part of the final paper over the course of the year. The background reading you do in the early part of the year will form the basis for your *Introduction*. There is no reason not to start working on your *Introduction* as soon as you can. Similarly, you could write a draft version of your *Method* section once you have worked out how the experiment will be run. Finally, you should consider preparing a draft outline of your *Results* section. This could simply be a list of each piece of data and analysis result, including templates of the data tables and figures you might use. If you know what kinds of data and analysis should be in your *Results* section when you write it up, then you can check to see if you have designed your experiment so that all of that information will be available. It is also useful to make up hypothetical data figures based on your experimental predictions; this will make it easier to interpret your results when you get them. Sometimes, even though you seem to have designed an experiment properly, you may realize when you come to plot up the data that you can't interpret them in the way you had hoped.

Of course, you will be making changes to your paper as you proceed. You will have to read more; you may need to modify your procedure; and you will have to write your *Discussion* section later on.

A FINAL WORD

Something else that may distinguish a thesis from a lab report or other kind of research paper is that you may be expected to submit a draft version to your adviser, who will then provide you with suggestions for changes. Don't be discouraged if the comments are critical; very few initial drafts are perfect. Instead, take advantage of the help your adviser can give with respect to the write-up.

chapter 9

PRESENTATION OF DATA

As we noted in Chapter 7, the *Results* section is the heart of an experimental paper. If you don't present your results clearly, the lab report will not be worth reading. If you have done anything more than the simplest experiment, you will find that you have to choose between many possible ways of organizing and presenting your results. For example, you will have to decide whether to use a figure or a table. If you choose to prepare a figure, you will have to decide whether to make it a bar chart or a graph. If you decide to make it a graph, you must decide how you will arrange the axes. And this is just the beginning! With experience, these decisions will become much easier to make, but at first you will need to spend some time thinking about presentation.

ORGANIZING A RESULTS SECTION

A set of experimental data consists of two main elements: the raw data you have collected and the results of any analysis you've chosen to do. A good *Results* section should contain both a summary of the data and a report of the analyses.

It is not unusual to see the *Results* section of a psychology report begin along these lines: "An analysis of variance was done on the data, and there was a significant main effect for X." This is not very informative. The *Results* section is where you begin trying to convince the reader of the validity of your work. Although you should not be interpreting your data in this section, you should be presenting them in such a way that your reader is directed towards the same conclusions that you plan to draw. As a general rule, you should start with either a statement of the question you were asking or a statement of your experimental hypothesis, and you should follow this with a brief summary of your main findings. Your guiding principle should be description first, then analysis. You should then lead your reader through the data, one level of analysis at a time, providing enough information to justify the conclusions that you will make in the *Discussion* section. Deal with the main findings first before going on to the secondary results. Keep in mind also that you must present all the important results, whether or not they support your hypothesis. Never neglect the "inconvenient" finding—your reader probably won't.

SUMMARIZING THE MAIN FINDINGS

Obviously, for most of your experiments, you cannot present all of your raw data. When you are measuring some characteristic of living organisms, there will almost always be variability across individuals, or even within individuals over successive measurements. For this reason it is usual to make repeated measurements of the same thing and take an average. Typically, you would begin by reporting your group means, together with some measure of variability. Although there are a number of different ways of describing the variability in a set of data, the most common are the standard deviation (*SD*) or the standard error of the mean (*SE*). (If you are not familiar with these terms you should consult a text on basic statistics.) For example, in an experiment that examined the effect of cartoon content on the number of aggressive acts in children (assuming you had ethics approval for such a study), you might summarize your results this way:

> Children who watched violent cartoons engaged in a larger number of aggressive acts (*M* = 7.3, *SD* = 1.2) than those who watched non-violent cartoons (*M* = 4.1, *SD* = 1.9).

A brief summary like this is appropriate if you have only one or two variables to deal with. But what if your experiment had been more complex? Let's say again that you were asking questions about TV violence and aggression in children. In this case, however, you wanted to know whether cartoons have different effects from violent live-action shows, and if the effects of these shows are different in boys and girls. Now you have eight means to report rather than just two. At this point you might want to consider using a data table or a figure.

DATA TABLES

If you can make your data summary clearer by using a table, by all means do so. However, it's essential that any tables you use be well organized and clear. Don't overload them with data, and be sure to plan the layout carefully. For the example above you might do something along the lines of Table 9.1.

With a table like this, the reader can see the results at a glance. A good table should be self-explanatory. Nevertheless, it's important to refer to each table in the text and point out the most important aspects, without repeating everything the table contains. For example:

> Table 9.1 shows that exposure to either cartoon or live-action TV violence increases the number of aggressive acts by children. This table also shows that girls exposed to live-action violence appear to become most aggressive.

TABLE 9.1   NUMBER OF VIOLENT ACTS BY CHILDREN
            FOLLOWING EXPOSURE TO TV VIOLENCE

	Cartoons		Live action	
	Violent	Non-violent	Violent	Non-violent
Group	M (SD)	M (SD)	M (SD)	M (SD)
Boys	6.7[a] (2.1)	4.3 (2.8)	6.4 (1.9)	4.7 (1.6)
Girls	5.4 (1.5)	3.8 (1.2)	7.1 (3.3)	4.0 (1.4)

Note. The means represent the number of observed aggressive actions during a half-hour observation period.
[a]One child in this group did not complete the whole experiment and was excluded from the analysis.

PREPARING TABLES

The tables in this chapter were prepared according to APA guidelines (American Psychological Association, 2001). Among the most important points to remember are these:

- Double-space everything in the table, including the titles and any notes at the bottom.
- Give each table a brief, informative heading.
- Do not use any vertical lines anywhere in the table. If necessary, use extra spacing between columns to make the table easier to read.
- Make sure that each column has a heading.
- Explain any abbreviations or special symbols you use in a note.
- If you have made statistical comparisons between items in the table, be sure to identify them with asterisks and give the significance levels in a note.
- If you do use notes to clarify information in the paper, they should be listed in this order:

  1. *general notes*, which provide information relating to the table as a whole;
  2. *specific notes*, which refer to individual cells in the table (for example, to mention that a subject dropped out of a group), identified by superscript, lower-case letters; and
  3. *probability notes*, which indicate the significance levels used.

These technical requirements aside, you should also ask yourself several questions:

- Do I have a brief, informative title on each table?
- Do the data in the table complement—rather than duplicate—what I have written in the text? (You should avoid repetition as much as possible.)
- Are the tables numbered sequentially throughout the paper?
- Is the style of the tables consistent throughout?
- Have I referred to the table in the text?

DATA FIGURES

The ability to represent data clearly in a suitable graphical form is a skill well worth learning. At the very least, you should be aware of the major kinds of graphs that you might use in a research project and the advantages of each. When you are preparing a figure, think about style as well as content; the most successful figure is often the simplest.

LINE GRAPHS

Line graphs, in their simplest form, plot the value of a dependent variable (on the vertical, or y-axis) against changes made in the independent variable (on the horizontal, or x-axis). They are particularly well suited to those cases where you want to show a continuous change in the value of a variable under different experimental conditions.

SCATTERGRAMS

Scattergrams are very useful when you want to express a correlation between two variables. When you plot a scattergram, each data point represents a measurement or score made under two conditions. If you are trying to make predictions about one score on the basis of another, then you would use the abscissa, or x-axis, to represent the independent (predictor) variable, and the ordinate, or y-axis, for the dependent (predicted) variable. The scatter of the points on the graph will provide the reader with a sense of the *variability* of the data—in other words, the strength of the relationship. You can also quantify the trend by calculating the best-fitting regression line.

BAR GRAPHS

Bar graphs consist of bars whose lengths are proportional to the value of the variables you are displaying. They are best suited to situations in which you want to compare the results of a limited number of experimental conditions—usually not more than two or three. If you include more than that, the graph becomes more difficult to interpret. Although there are several ways of plotting bar graphs, the most important thing to keep in mind is that you should put the conditions you want to compare most directly next to each other.

HISTOGRAMS AND AREA CHARTS

Histograms and area charts are not used extensively in psychology reports, but they are useful if you want to show how many participants received a particular score on a test or how many of the participants fell into each of a series of different categories. The conventional histogram will have multiple bars, each representing a subset of the data you have collected. An area chart is simply a histogram that plots only the mid-points of the bars and therefore looks more like a line graph.

PIE CHARTS

Pie charts present proportions graphically like slices of pie on a plate. They are not often used in scientific papers, although they are very common in media presentations or reports directed to the general public. The results of survey research are sometimes displayed in pie charts.

SOME EXAMPLES

Let's reconsider the data in Table 9.1. You could present these data very easily in the form of a bar graph, as shown in Figure 9.1. In this figure we have chosen to make two separate bar graphs, one for cartoons and one for live action. We have also chosen to juxtapose the *violent* and *non-violent* bars for the closest comparisons. If the emphasis of the study had been on differences in boys' and girls' behaviour, we might have put the *Boys* and *Girls* columns next to each other. Depending on what aspects of your data you want to emphasize, you will have to decide which is the best way to arrange your figure.

In this example, there is no strong reason for choosing either a figure or a table. However, if you have a lot of data and the point they make might be unclear even with a large table, then a figure is your best choice. As with tables, be sure to refer to each figure in the text of your report and mention the major points that it illustrates.

Now let's suppose that instead of showing all the children the same violent scenes, you varied the number of violent acts in the films in order to see whether quantity has an effect on aggressive behaviour. One way of reporting such data is in the form of a table, as shown in Table 9.2.

FIGURE 9.1   MEAN NUMBER (SE) OF AGGRESSIVE ACTS BY
CHILDREN FOLLOWING EXPOSURE TO VIOLENT OR
NON-VIOLENT TV EPISODES

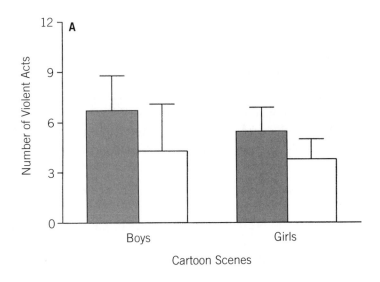

Cartoon Scenes

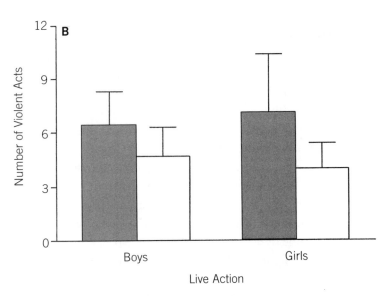

Live Action

Figure 9.1 This figure shows the data from Table 9.1 in the form of bar graphs. The bars represent the mean number of aggressive acts by children following exposure to violent or non-violent TV episodes. Panel A shows the cartoon scenes; Panel B, the live-action scenes. Dark bars: violent scenes; light bars: non-violent scenes. The vertical lines represent the standard deviations that are shown in the table.

TABLE 9.2 NUMBER OF AGGRESSIVE ACTS BY CHILDREN FOLLOWING EXPOSURE TO TV VIOLENCE

	Number of violent incidents in TV film							
Group	0	1	2	3	4	5	6	7
Boys	.3	1.4	2.3	2.6	3.3	5.6	5.8	5.7
Girls	.2	1.1	2.5	3.8	4.1	3.2	2.4	1.2

These numbers suggest that there is a tendency for boys to show increasing aggression with increasing exposure, whereas girls show an increase followed by a decrease. A data table of this sort is an acceptable way of presenting your results, although if you tried to include a measure of variability, the table would start to become cluttered. Contrast this table with a figure showing the same set of data. In Figure 9.2 you can see at a glance how differently the two groups react.

FIGURE 9.2 MEAN NUMBER OF AGGRESSIVE ACTS FOR BOYS AND GIRLS AS A FUNCTION OF THE NUMBER OF VIOLENT TV SCENES TO WHICH THEY WERE EXPOSED

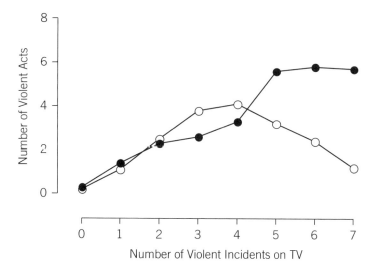

Figure 9.2 This figure shows the data from Table 9.2 in the form of a line graph. Filled symbols represent the boys, open symbols the girls. Note how much easier it is to see how the number of violent acts by the girls increases then decreases.

The actual numbers are probably less important than the *pattern* of results shown in the figure. You could take this analysis further and look at the strength of the relationship between amount of exposure and aggression by creating a scattergram, as shown in Figure 9.3. By plotting all of the points for each individual subject you can see if this is a very strong trend or just a weak relationship.

Scattergrams are effective when you want to show a correlation between two variables—such as violent TV scenes and aggressive acts—but they aren't suitable for illustrating data that fall into a number of categories. Suppose, for instance, that you wanted to find out how students *really* spent their time while they were "working." You provide a group of students with a diary to keep track of the amount of time they spend in various activities. There are several possible ways of presenting your data. A line graph would not be appropriate for these data, but there are alternatives. You could make a table, listing the percentage of time students spent doing different things; or you could draw a bar

FIGURE 9.3

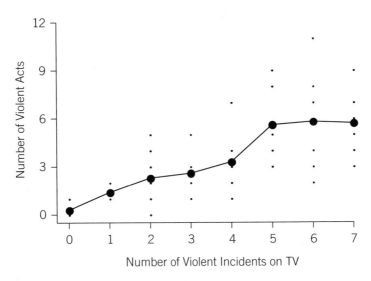

Number of Violent Acts

Number of Violent Incidents on TV

Figure 9.3 The data in Table 9.2 and Figure 9.2 represent means. This figure shows how you could represent the data from individual participants as a scattergram. For each of the experimental conditions the number of violent acts for six boys is shown. Because several had the same score, the points overlap. The filled circles and the connecting line represent the mean data. By using a scattergram, you can get a better sense of the degree of relationship between two variables. You could also go one step further and calculate a regression equation so that you could fit a regression line to the data.

FIGURE 9.4 PERCENTAGE OF TIME STUDENTS SPENT IN
         DIFFERENT ACTIVITIES DURING AN EVENING
         SESSION WHEN THEY WERE "WORKING"

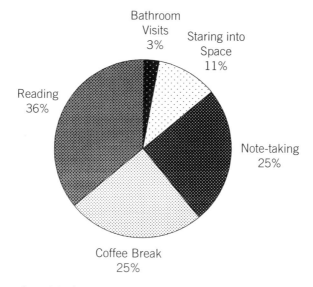

Figure 9.4 When you have data that can be broken into percentages or proportions, a pie chart gives the reader intuitive sense of the distribution.

graph to make the point more graphically. In this case, however, you could probably make the greatest impression by presenting a pie chart, as in Figure 9.4. Here it is easy to see that your sample of students actually spent less than 50 per cent of their "work time" doing things that might be construed as work.

As you can see, there are many ways to present your data graphically. Figures that are informative and well presented can make the difference between a mediocre grade and a good one.

PREPARING FIGURES

Most of the figures you present will be graphs. Many spreadsheet programs, such as Microsoft Excel, have the graphing capabilities to produce graphs that should be quite acceptable for a term paper. However, because these programs were originally designed for preparing business presentations, the default versions of the graphs they produce are probably not what you would want to use in your paper. Nevertheless, with a little reorganization, you can produce graphs that conform to the format typically used in your discipline. There are also a number of commercially available packages that will allow you to create virtually any kind of graph that you might want to use. If you are working on your thesis, it is very likely that your adviser will have a program that you can use. You should also check to see if your institution has a site licence for one of these packages. Two of the most popular are *SigmaPlot*™ and *GraphPad* Prism™. If you don't have access to a computer graphics package, then you will need to draw your figures by hand. Whether you are preparing your graphs by hand or on a computer, the following APA guidelines will be helpful:

- Your figures should be as neat as possible.
- If you are drawing a figure by hand, use a ruler and black ink. You might consider drawing a rough version on graph paper and then tracing it onto plain paper.
- The independent variable (the one you have manipulated) should be on the horizontal axis, and the dependent variable (the one you measure) on the vertical axis.
- The vertical axis should be about 75 per cent of the length of the horizontal one.
- Use large and distinctive symbols, with different symbols for each line on the graph. If you have multiple graphs with the same data categories, be consistent and use the same symbols for each category.
- Label the axes clearly. Start each main word with a capital and run the axis label parallel to the axis. Always include the unit of measurement on the axis label.
- Show only essential detail. A cluttered graph is difficult to read.
- Include a legend describing the figure and indicating what the different symbols represent.

MAKING YOUR FIGURES MORE INFORMATIVE

The suggestions we have made above will give you the tools to prepare simple, informative graphs. You should be aware, though, that a graph can be modified in a variety of ways to provide much more information to the reader.

CURVE-FITTING

We have already mentioned the possibility of calculating a regression equation and plotting the best-fitting straight line onto your scattergram. If your data are more complex, then you may want to use more sophisticated curve-fitting programs that will allow you to fit and plot other functions onto your data. For instance, if the measurements increase and then decrease, you might need to calculate a non-linear regression equation, or you might find that your measurements are best described by an exponential function. In some cases, particularly when you are dealing with biological data, these functions will give you additional information about the underlying mechanisms that are responsible for the pattern of results you obtain.

MEASURES OF VARIABILITY

Another, more common addition to both bar and line graphs is some indication of the variability within the data. Depending on the kind of data you have and the point you are trying to illustrate, this variability might be the standard deviation, or the standard error of the mean, or the 95 per cent confidence limits. Figure 9.5 shows a line graph that contains standard error bars. Notice that when two lines are close together, as they are in this figure, the graph can be made neater by plotting the error bars for each in opposite directions.

FIGURE 9.5

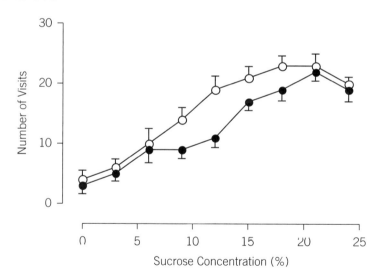

Figure 9.5 This figure shows a set of fictitious data on the feeding behaviour of humming birds. The graph shows the mean number of visits to a bird feeder, plotted as a function of the concentration of sucrose in the feeding solution. The open symbols represent birds that were nesting, the filled symbols, birds that were not nesting. The bar represents one standard error of the mean. Standard error bars are a useful way to indicate the amount of variability in the data. The standard error is an estimate, based on the standard deviation, of the amount of variability within a whole population, rather than just the sample you tested. It forms the basis for some statistical tests.

USE OF DIFFERENT SCALES

Sometimes you may have data that cover a very wide range of values. This is not unusual if you are working in some areas of biology or in sensory psychology. If you plot these data on a regular graph, any differences that exist at low values might become invisible at that scale. One way to get around this problem is to use a logarithmic scale. What you are doing when you create a log scale is

FIGURE 9.6

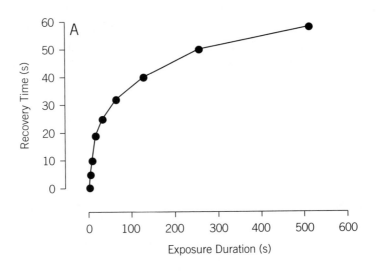

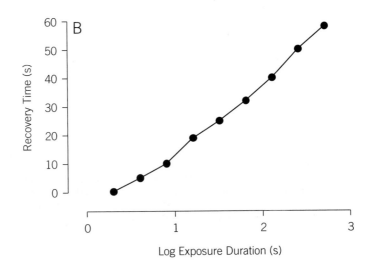

FIGURE 9.6   CONTINUED

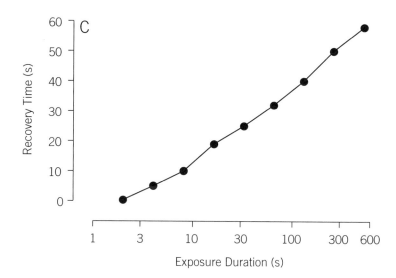

Figure 9.6 This figure shows a set of fictitious data on the recovery of visual sensitivity following exposure to a very bright light. The same data are plotted in two different ways. In Panel A, standard linear axes are used, and they indicate that recovery is very fast following brief exposure but is much slower after long exposures. The same data are plotted in Panel B, but this time the exposure durations, plotted along the x-axis, have been converted to base-10 logarithms. An alternative way to plot the x-axis would be to keep the logarithmic scale but to use the original (antilog) numbers, as shown in Panel C.

converting your data values into their logarithms and plotting these on your graph. The effect of this is to expand the low end of the scale and compress the high end. Figure 9.6 shows you the same set of data plotted on both linear and log x-axes. Once the values are converted to logs it is much easier to see how the details change.

There are other refinements you could make to your figures. But even if you follow only the suggestions here, your figures will be much better than your instructor is accustomed to seeing in student lab reports.

TABLES OR FIGURES?

When deciding whether to use a table or a figure to present your data, you must always consider which will be the most effective way of getting your point across. Tables have the advantage of being more exact than figures because they provide precise numerical data. On the other hand, figures often give a more compelling impression of the overall pattern of results. Your decision to make a table rather than a graph should be based on which you think will convey the information most effectively.

In general, a table is a good choice if you have several sets of numbers that could get buried if you just listed them in the text. However, if you have collected data for a number of conditions that vary systematically, then a graphical representation is by far the best way to look for, or express, a functional relationship. As a general rule, a figure should go beyond what you can provide in numerical or textual form, so if you think that a graph would provide no more information than a table, it is probably best to go with the table. In some cases the choice between a table and a figure will come down to nothing more than personal style or preference.

PRESENTING YOUR STATISTICAL ANALYSES

For the most part, your figures and tables are descriptive: they show what the results are and serve as accompaniments to the text. The next step is to give an account of the statistical tests you have done.

If the analysis is straightforward, you can report it directly in the text, giving enough information for the reader to confirm that the analysis has been done correctly. The exact information will vary with the test, but it may include the value of the test statistic, the degrees of freedom, and the level of statistical significance.

If your analysis is a complicated one—for example, a multifactorial analysis of variance, or an analysis involving a series of correlations—you might consider using a statistical table showing all the details of your analysis. As you become more experienced at presenting data, it will become easier to decide what to include in the text or in tables. Initially, though, you should refer to the *American Psychological Association Publication Manual* (American Psychological Association, 2001) or ask your instructor what is required.

The goal of many of your experiments will be to test a hypothesis. Even if you are asking a question for which you do not have a specific prediction, you may set your experiment up formally as a test of a statistical hypothesis. For example, the question "What is the effect of different drugs on the rate of learning in rats?" could be rephrased as a statement: "Drug A produces significantly better learning than drug B." You could then test that hypothesis using the appropriate statistical tests.

In your analysis you should present the actual statistical values followed by a brief statement of what they mean. Although you may know exactly what your

F-ratios imply, your reader may not. For example, in the experiment testing the effect of different drugs on the rate of learning, you might have analyzed your data using an analysis of variance. You could describe the results of this analysis by saying,

> The analysis of variance showed a significant difference in the effectiveness of the two drugs ($F(1, 15) = 14.5$, $p<.01$). The rats that were given drug A took fewer trials to learn the task than those that received drug B.

Finally, if you have performed several different analyses on your data, you should start by reporting the overall analysis and then move on to the secondary analyses. Remember, you are leading the reader through your paper, and the better your organization, the clearer it will be.

REFERENCE

American Psychological Association. (2001). *Publication manual of the American Psychological Association* (5th ed.). Washington, DC: American Psychological Association.

chApter 10

APA Editorial Style

The last five chapters have offered you guidelines on how to prepare several different kinds of written assignment. The main goal of these chapters was to get you to consider how to organize your thoughts and your writing in order to present your story clearly. Now it is time to think about the style of that presentation. In this context, the term "style" has two different meanings. The first refers to the manner in which we express our ideas: the words we choose, the length of sentences, and so on; this expressive style is discussed in Chapter 14. The second kind of style is editorial, and concerns the physical arrangement of the words on the paper. For anyone working in psychology, the definitive style reference is the *Publication Manual of the American Psychological Association*, now in its fifth edition (American Psychological Association, 2001). Although the *APA Manual* is directed mainly to those who publish in journals sponsored by the Association, its rules on format and style have been adopted by several academic disciplines as a guide for publishing articles and books, as well as for preparing student papers and theses; these guidelines can be applied to most scientific writing, even in disciplines that do not explicitly use APA style.

RULES OF PRESENTATION

The *APA Manual* devotes almost 200 pages to APA editorial style, covering everything from how the major organizational headings should appear, to when a comma should precede the word "and." Obviously we cannot describe all the APA guidelines here; we will give you only the ones that are most important for undergraduate papers. For more details, refer to the manual itself.

PAPER FORMAT
The following guidelines apply to how the individual pages of your report or essay should appear:

- Use a single side of each page and double-space the entire paper, including the references.

- Number each page, including the first, with Arabic numerals in the top, right-hand corner. It's also good practice to use a *page header*—a short version of the title—at the top of each page, either immediately above or just to the left of the page number; that way, if any of the pages gets separated it will be easy to identify.
- Always leave generous margins—at least an inch (2.54 cm) on all sides—so that your instructor can add comments.
- In a paper submitted for publication, you must always place tables and figures at the end of the manuscript; however, for a student paper, you may incorporate the tables and figures into the text, unless your instructor tells you otherwise. Normally you should place these on a separate sheet, immediately following the page where they are first referred to. Depending on the rules for your course, you may place the figure legend either under the figure itself or on a separate page, facing the figure.

HEADINGS AND SUBHEADINGS

When you are writing a library research paper it is sometimes useful to break down the paper into several sections, particularly if you have organized your paper according to an outline, as discussed in Chapter 6. When you are writing a lab report, headings and subheadings are essential. APA style permits up to five levels of headings and subheadings:

LEVEL FIVE
(centred, upper-case)

Level One
(centred, upper- and lower-case)

Level Two
(centred, upper- and lower-case, italicized)

Level Three
(flush left, upper- and lower-case, italicized)

 Level four. (indented paragraph heading, italicized, lower-case after initial letter, ending with a period and leading directly into paragraph)

You will not often need to use all of these levels; in most cases, two will be adequate. For example:

	Method	(Level 1)
Apparatus and Procedure		(Level 3)

If you need a third level, add *Level 4*:

	Method	(Level 1)
Subjects		(Level 3)
Control group.		(Level 4)

If you had a series of experiments to describe, you might need to rearrange the headings:

	EXPERIMENT 1	(Level 5)
	Method	(Level 1)
Subjects		(Level 3)
Control Group.		(Level 4)

The most important thing to remember is that headings should follow a logical and consistent pattern. Always start with Level 1 (except in a five-level paper) and work down. You may not use every level in each section, but you should ensure that equivalent sections or sections of equal importance have the same level of heading.

UNITS OF MEASUREMENT

All APA journals require that any measurements be expressed in metric units, using, in most cases, the International System of Units (SI). If you did not make your original measurements in metric units, you should report the metric equivalents for these measurements. For clarity, you may want to report both: for example, "The platform was placed 15 in. (38 cm) above the landing surface." You should also be sure to use the appropriate SI abbreviations for the units (*see page 229*).

NUMBERS

In general, you should use figures to express all numbers greater than nine (10 and above) and numbers below 10 that are used in comparisons with numbers above 10: for example, "In a recent study, 5 out of 20 subjects reported that . . .". However, never begin a sentence with a figure, as in, "5 out of 20 subjects in a recent study . . ."; in this case, you must either write out the number and all related numbers in the sentence ("Five out of twenty subjects in a recent study . . .") or recast the sentence so that it does not begin with a number.

Other items that should always be expressed in figures include all numbers relating to statistical and mathematical functions and measurement; times, dates, and ages; numbers of subjects in an experiment; and numbers that describe part of a series. In most other circumstances—for example, when expressing common fractions or beginning a sentence, title, or text heading with a number—you should use words. When in doubt, go with the form that makes your statement easiest to read.

QUOTATIONS

We have already stressed the importance of acknowledging any quotations that you use. You should also be aware of how to present the quotations in the text:

- **Short quotations.** If the quotation contains fewer than 40 words, enclose it in double quotation marks (". . .") and include it as part of the text.
- **Long quotations.** A longer quotation is set off from the main body of the text by indenting the whole quotation five to seven spaces from the left margin. Although APA style does not require it, you might ask your instructor whether you should also indent on the right. The quotation should not begin with a paragraph indent, but additional paragraphs should be indented from the new left margin. You do not need quotation marks for this kind of block quotation.
- **Quotes within quotes.** If the passage you are quoting includes material already in quotation marks, make these single if the main quotation is in running text and double if it is set off as a block quotation. For more information on using quotation marks, see Chapter 17.
- **Changes to the source.** If you change any words within the quotation—for instance, to maintain grammatical sense—the non-original words should be enclosed in [brackets], not (parentheses). To create emphasis, use italics followed immediately by [italics added]. When you omit part of a quotation, indicate the missing section with an *ellipsis*—three spaced periods (. . .). If the omission is between sentences, include the original period (without a space) before the ellipsis. For more information on ellipses, see Chapter 17.
- **Acknowledging sources.** You should acknowledge the source of the quotation in the text, giving the page number(s), as well as the author and the year of publication. The complete reference should be

contained in the reference list. Chapter 11 describes the format for references in APA and other styles.

APA-STYLE HELPER

If you are likely to be writing a lot of papers that will use APA format, there is a useful tool to help you create APA-style documents. This is the *APA-Style Helper* (http://www.apa.org/software/), published by the American Psychological Association. It is a software package that works in conjunction with Microsoft Word to help in the preparation of APA-formatted reports. It will help you organize your document using the appropriate APA sections, and it will allow you to build a database of references that will be formatted appropriately. As you write your paper you can insert citations and create a reference list that will be appended to your report automatically. The reference database that you create will be saved so that you can include the same reference in another paper without having to retype it.

REFERENCE

American Psychological Association. (2001). *Publication manual of the American Psychological Association* (5th ed.). Washington, DC: American Psychological Association.

chapter 11

DOCUMENTATION IN THE SCIENCES

There is a huge amount of variation in accepted formats for referencing your sources in academic writing. In psychology there is a uniform style for referring to the literature you have cited in your work. However, documentation methods differ quite widely among the other sciences, and even within some disciplines. As a result, you should always check with your instructor to make sure you are following the preferred practice for the course you are taking. Some instructors are very strict about citation practices; others will give you more freedom. The two most important things to remember are that your citations must be accurate and whatever style you use must be consistent.

CITATIONS

Documentation in scientific writing differs from that in the humanities in two important ways:

1. Instead of footnotes, brief references (called *citations*) are included in the text wherever reference to another person's work is made.
2. At the end of the paper, instead of a bibliography, there is a section called *References*, which lists the full publication details for only those works referred to directly in the text; other works that have been consulted but not actually cited in the paper are not listed.

There are two main styles of citation in scientific writing: *alphabetical* and *numerical* (or *consecutive*). Within each of these styles there are several ways of citing the references in the text. In addition, the format for listing works in the *References* section varies widely. This chapter will show you the most common citation formats and suggest journals that you may use as models. Although this chapter will concentrate on APA and CSE reference styles, you should bear in mind that there are alternatives you may wish—or be instructed—to use.

ALPHABETICAL CITATIONS

The most widely used alphabetical system is the one defined in the APA's *Publication Manual* (American Psychological Association, 2001). Virtually all psychological publications, as well as many in the social sciences and other academic disciplines, use this format. A variation of the APA alphabetical system is outlined in the style manual published by the Council of Science Editors (2006), which is used by many writers in the life sciences and medicine. (The *CSE Manual* also describes a system of numerical citations, which is explained later in this chapter.) The CSE's alphabetical system is quite similar to that of the APA but differs from it in that it makes much less use of punctuation such as periods and commas.

The APA and CSE methods of alphabetical citations are two models you should consider using. Be aware that in addition to the fact that these two models differ in places, each one is constantly undergoing revision to keep up with new ways of doing research. If you are uncertain of how to document a particular source, you should refer to the *APA Manual* or the *CSE Manual*. If you do not have a copy available when you are writing your paper, there a number of websites that can provide you with information. One that we have found helpful for APA style is http://webster.commnet.edu/apa/apa_intro.htm. You may also wish to consult the APA's own website to review the latest guidelines on citing electronic sources: http://www.apastyle.org/elecref.html. For information on CSE style, you can try http://www.councilscienceeditors.org/publications/citing_internet.cfm.

GUIDELINES FOR USING ALPHABETICAL CITATIONS

1. A parenthetical citation including the author's surname and the year of publication is inserted into the text at the most appropriate point, often at the end of the sentence in which you refer to the work:

    APA   In a study of children of alcoholics (Birchmore, 1999) . . .

    CSE   In a study of children of alcoholics (Birchmore 1999) . . .

    Note that the CSE does not use a comma to separate the author's name from the date.

2. If reference to the author has already been made within the text, it can be omitted from the citation:

    Birchmore (1999) studied the children of alcoholics . . .

    In 1999, Birchmore reported . . .

3. If there are two authors, include both names every time you cite the reference in the text. The APA uses an ampersand (&) when the names are in parentheses but "and" in the running text:

APA   Earlier research showed that neonates did not differentiate between small frequency differences (Leventhal & Lipsitt, 1964).

CSE   Earlier research showed that neonates did not differentiate between small frequency differences (Leventhal and Lipsitt 1964).

APA, CSE   However, Morrongiello and Clifton (1984) have found that neonates will orient better to high- than to low-frequency sounds.

If there are three, four, or five authors, the APA style cites all the names when the reference first occurs, and afterwards cites only the first author, followed by "et al." For six or more authors, the text citation should contain only the name of the first author, followed by "et al.":

APA   Pola, Wyatt, and Lustgarten (2001) studied eye movements.

then   Pola et al. (2001) found that subjects suppressed movements.

Research found that subjects suppressed movements (Pola et al., 2001).

If there are three or more authors, the CSE style cites the first author's name followed by "and others":

CSE   Pola and others (2001) found that subjects suppressed movements.

Research found that subjects suppressed movements (Pola and others 2001).

Note that the year should be included the first time a work is cited in a paragraph but not in subsequent citations within that paragraph (unless you mention more than one work by the same author).

4. If the page reference is not important, all you need to include in the citation is the author's name and the year. If, however, you are referring to a particular part of a source, you must give the specific location. Always give the page number if you quote a source directly. Note that the APA uses "p." (with a period) for a single page number and "pp." for several pages, whereas the CSE uses "p" (without a period) regardless of whether the reference is to one or several pages:

APA   (Cormack, Stevenson, & Schor, 1994, pp. 2601–2602)
(Schmidt, 1998, fig. 2)
(Choudhry, 2001, table 3.1)
(Li & Knowles, 2002, chap. 8)

CSE   (Cormack, Stevenson, and Schor 1994, p 2601–2)
(Schmidt 1998, fig 2)
(Choudhry 2001, table 3.1)
(Li and Knowles 2002, chap 8)

5. If you are citing several papers written by a single author, list the surname once with the dates following chronologically and separated by commas:

APA   (Smith, 1980, 1996, 2000, in press)

CSE   (Smith 1980, 1996, 2000, in press)

Several papers published by the same author in the same year are distinguished with lower-case letters (a, b, c, etc.) after the year and separated by commas:

APA   (Smith, 2003a, 2003b)

CSE   (Smith 2003a, 2003b)

Be sure to use the same letter code in the list of references at the end of your work.

6. Several papers by different authors can be included in the same citation, separated by semi-colons. In APA style, references are cited in alphabetical order according to authors' surnames. In CSE style, references are listed in chronological order; references to works published in the same year are sequenced alphabetically:

APA   (Aaron, 2001; Diaz & Quan, 1980; Hicks et. al., 1992; Schiller, 1980; Zuk, 1932)

CSE   (Zuk 1932; Diaz and Quan 1980; Schiller 1980; Hicks and others 1992; Aaron 2001)

7. If authors have the same last names, use initials to distinguish them:

APA   (Roy, B.G., 1999; Roy, P.E., 1999)

CSE   (Roy BG 1999; Roy PE 1999)

8. If the work you are referencing has no known or declared author, cite the first few words of the reference list entry:

APA ("Substance abuse becoming a problem," 1999) [In this case, the full title of the article might be "Substance abuse becoming a problem among seniors, study shows."]

CSE (Anonymous 1999)

9. Sometimes, if you are unable to locate an original work, you may have to rely on a description in a secondary source. In this case you should cite the secondary source that you have read and include only that work in the *References* section:

APA According to Wilcox and Wolf (1999, as cited in Symons & Pearson, 2003, p. 16) . . .

CSE According to Wilcox and Wolf (1999, cited in Symons and Pearson 2003, p 16) . . .

(Wilcox and Wolf 1999, cited in Symons and Pearson 2003, p 16)

10. Machine-readable data files are sources of information that exist only in electronic form. These might include census data or other large databases of information. In the text, simply cite authorship and the date:

APA (Institute for Survey Research, 2001)

CSE (Institute for Survey Research 2001)

NUMERICAL CITATIONS

An alternative to the author–date citation used in APA style is the numerical citation. Several biological and medical journals use this style, in which the source is cited as a number that corresponds to the reference listed in the *References* section. The cited references may be placed in brackets or parentheses, or set as superscripts in the appropriate places, depending on which variation of the citation-sequence system you follow.

Some journals (such as *Nature* and *Science*) list the references in the *References* section in the order of their first citation in the paper; others use numeric citations in the text but arrange the reference list in alphabetical order. In the latter case, the citation numbers in the text cannot be assigned until all the references have been collated: if you take out or add any references, you will need to change the citation numbers.

The following are the CSE's guidelines for using in-text citation numbers. Remember that variations of the numerical system exist, so you should always find out which style is preferred by your instructor or department:

1. Superscript numbers in the text correspond to numbered references in the *References* section. These references are numbered according to the order in which they are first cited in the text. Thus, the first reference used in the text will be [1]; the next new reference cited will be [2]; and so on. Once a source has been assigned a number, it is referred to by that number wherever it appears in the text:

   > Labossière's groundbreaking study[1] was challenged first by Gormon[2] and later by Huang[3]. In fact, Gormon gained considerable notoriety for her harsh criticism[2] of Labossière's interpretation of the results[1].

2. If you are referring to several sources in one reference, separate the numbers with commas and no spaces. When citing a sequence of three or more citation numbers (i.e., 7, 8, 9), use only the first and last number in the sequence, separated with a hyphen:

   > This phenomenon, which has been described by several authors[2,3,5,7-9], is . . .

3. When citing material that you have not read yourself but have seen cited by others, give the citation number for the original followed, in parentheses, by "cited in" and the citation number for the source in which you found the material:

   > According to Wilcox and Wolf[18(cited in 19)], . . .

FORMAT OF THE REFERENCE SECTION

The reference list is an extremely important part of your paper. It should contain every work that you have cited in the text—and none that you have not cited. (To include uncited references in your *References* section is fabrication.)

The precise format of the reference list varies among disciplines and journals. Here again we concentrate on APA and CSE models, but most of the general rules about the organization of the reference list are similar across disciplines. The differences occur primarily in the placement of publication date, presentation of authors' names, use of abbreviations, and punctuation. If your discipline does not use APA or CSE style explicitly, ask your instructor which journal to use as a model.

APA GUIDELINES

The most important APA requirements for the reference list are these:

1. References are typed with the first line flush left and subsequent lines indented, as shown in the examples below. It is permissible to use the paragraph indent if the hanging indent is difficult to accomplish with the author's word-processing software.
2. Arrange the references alphabetically beginning with the first author's surname, followed by the initials. Alphabetize references for works with no known or listed author according to the first important word of the title.
3. The date of publication follows immediately after the authors' names.
4. When citing a work with several authors, list the first six authors' names, surnames first, then initials, followed by "et al." (In some biological journals, only the first three authors are listed. In some styles, only the first author's name and initials are reversed.)
5. For works published in the same year by a single author, add the letter suffix (1992a, 1992b, etc.), as you did in the text.
6. When citing more than one work by the same first author but with different subsequent authors, list the references alphabetically first (by surnames of the subsequent authors), then chronologically if more than one reference has the same combination of authors.
7. Give the full title of the article, chapter, or book. (Some journals, such as *Nature* and *Science*, do not give titles, but it is unlikely that you will be expected to use this type of style. If your discipline's style permits abbreviations in journal titles, use the standard forms.)
8. Only proper nouns and the first word of the title and, if there is one, the subtitle are capitalized.
9. For journal references, always give the year, volume, and page numbers. (Some journal styles require only the first page to be listed. Again, it is unlikely that you will use this style.)

CSE GUIDELINES

The CSE has different guidelines for the *References* section, depending on whether you are using the author–date (alphabetical) system of citations or the citation-sequence (numerical) system. The two principal differences between the two styles are these:

1. If you are using the author–date citation system, your references should be listed *alphabetically* in the reference list. If you are using the citation-sequence system, your references are numbered *sequentially* and listed according to when they are first cited in the text of your paper.
2. In references corresponding to alphabetical citations, where the year of publication is used to identify the source, the year comes immediately after the authors' names. In references corresponding to numerical citations, the year is placed near the end of the entry.

Apart from these differences, the two styles of CSE references are essentially the same. In the examples below, you will notice that the CSE tends to favour fewer punctuation marks and more abbreviations than are used in APA style.

SAMPLE ENTRIES

The following examples of APA and CSE references cover the most common kinds of sources you will have to refer to. To cite other kinds of sources, check the detailed instructions in the *APA Manual* or the *CSE Manual*, or have a look at one of the websites mentioned earlier in this chapter.

ARTICLE BY ONE AUTHOR IN A JOURNAL

APA   Irwin, H.J. (2001). The relationship between dissociative tendencies and schizotypy: An artifact of childhood trauma? *Journal of Clinical Psychology, 57*, 331–342.

CSE   Irwin HJ. 2001. The relationship between dissociative tendencies and schizotypy: an artifact of childhood trauma? J Clin Psychol 57:331–42.

or   1. Irwin HJ. The relationship between dissociative tendencies and schizotypy: an artifact of childhood trauma? J Clin Psychol 2001;57:331–42.

ARTICLE BY TWO OR MORE AUTHORS IN A JOURNAL

APA   Bidlingmaier, S., & Snyder, M. (2004). Regulation of polarized growth initiation and termination cycles by the polarisome and Cdc42 regulators. *Journal of Cell Biology, 164*, 207–218.

CSE   Bidlingmaier S, Snyder M. 2004. Regulation of polarized growth initiation and termination cycles by the polarisome and Cdc42 regulators. J Cell Bio 164:207–18.

or   2. Bidlingmaier S, Snyder M. Regulation of polarized growth initiation and termination cycles by the polarisome and Cdc42 regulators. J Cell Bio 2004;164:207–18.

ARTICLE IN A MAGAZINE

If you are referring to an article in a magazine, you should also include both the volume number and the date shown on the issue—month only for monthlies such as *Scientific American*, month and day for weeklies:

APA   Stoye, J.P., & Coffin, J.M. (2000, February 17). A provirus put to work. *Nature, 403*, 717–719.

CSE   Stoye JP, Coffin JM. 2000 Feb 17. A provirus put to work. Nature 403:717–9.

or   3. Stoye JP, Coffin JM. A provirus put to work. Nature 2000 Feb 17; 403:717–9.

BOOKS

For books published in the United States, give the city and state of publication as well as the publisher; for the state, use the US Postal Service's two-letter abbreviations. For books published elsewhere give the province and country of publication. There are a few exceptions to these rules: major cities that are well-known publishing centres, such as Toronto, San Francisco, Amsterdam, or London, do not require further identification.

APA   Tomasello, M. (2003). *Constructing a language: A usage-based theory of language acquisition*. Cambridge, MA: Harvard University Press.

CSE   Tomasello M. 2003. Constructing a language: a usage-based theory of language acquisition. Cambridge (MA): Harvard Univ Pr. 388 p.

or   4. Tomasello M. Constructing a language: a usage-based theory of language acquisition. Cambridge (MA): Harvard Univ Pr; 2003. 388 p.

Note that the CSE includes the total number of pages as the last element of book entries.

BOOK WITH TWO OR MORE AUTHORS

APA   Peterson, L.M., & Russell, A.F. (2002). *Active and passive movement testing*. New York: McGraw-Hill.

CSE   Peterson LM, Russell AF. 2002. Active and passive movement testing. New York: McGraw-Hill. 418 p.

or   5. Peterson LM, Russell AF. Active and passive movement testing. New York: McGraw-Hill; 2002. 418 p.

BOOK WITH AN ORGANIZATION AS AUTHOR

APA   National Advisory Committee on Immunization. (2002). *Canadian Immunization Guide.* Ottawa: Canadian Medical Association.

CSE   National Advisory Committee on Immunization. 2002. Canadian immunization guide. Ottawa: Canadian Medical Assoc. 278 p.

or   6. National Advisory Committee on Immunization. Canadian immunization guide. Ottawa: Canadian Medical Assoc; 2002. 278 p.

BOOK WITH AN EDITION NUMBER
If the reference is a second or subsequent edition, be sure to include that information immediately following the title:

APA   Gleitman, H. (1995). *Psychology* (4th ed.). New York: Norton.

CSE   Gleitman H. 1995. Psychology. 4th ed. New York: Norton. 390 p.

or   7. Gleitman H. Psychology. 4th ed. New York: Norton; 1995. 390 p.

BOOK WITH A VOLUME NUMBER

APA   Musling, J.M., & Bornstein, R.F. (Eds.). (1998). *Empirical studies of psychoanalytic theories: Vol. 7. Empirical perspectives on the psychoanalytic unconscious.* Washington, DC: American Psychological Association.

CSE   Musling JM, Bornstein RF, editors. 1998. Empirical studies of psychoanalytic theories. Volume 7, Empirical perspectives on the psychoanalytic unconscious. Washington (DC): American Psychological Assoc. 291 p.

or   8. Musling JM, Bornstein RF, editors. Empirical studies of psychoanalytic theories. Volume 7, Empirical perspectives on the psychoanalytic unconscious. Washington (DC): American Psychological Assoc; 1998. 291 p.

BOOK WITH AN EDITOR

APA   Case-Smith, J. (Ed.). (1998). *Pediatric occupational therapy and early intervention* (2nd ed.). Boston: Butterworth-Heinemann.

CSE   Case-Smith J, editor. 1998. Pediatric occupational therapy and early intervention. 2nd ed. Boston: Butterworth-Heinemann. 324 p.

or   9. Case-Smith J, editor. Pediatric occupational therapy and early intervention. 2nd ed. Boston: Butterworth-Heinemann; 1998. 324 p.

CHAPTER IN AN EDITED BOOK

In the text you should cite the authors of the chapter you have read, and then give both the chapter title and the full book reference in the *References* section:

APA   Rogers, A.G. (1998). Understanding changes in girls' relationships and in ego development: Three studies in adolescent girls. In A. Westenberg, A. Blasi, & L.D. Cohen (Eds.), *Personality development: Theoretical, empirical, and clinical investigations of Loevinger's conception of ego development* (pp. 145–162). New Jersey: Erlbaum.

CSE   Rogers AG. 1998. Understanding changes in girls' relationships and in ego development: three studies in adolescent girls. In: Westenberg A, Blasi A, Cohen LD, editors. Personality development: theoretical, empirical, and clinical investigations of Loevinger's conception of ego development. New Jersey: Erlbaum. p 145–62.

or   10. Rogers AG. Understanding changes in girls' relationships and in ego development: three studies in adolescent girls. In: Westenberg A, Blasi A, Cohen LD, editors. Personality development: theoretical, empirical, and clinical investigations of Loevinger's conception of ego development. New Jersey: Erlbaum; 1998. p 145–62.

SIGNED NEWSPAPER ARTICLE

APA   Lawlor, A. (2001, June 26). A new kind of life line? *The Globe and Mail*, p. L3.

CSE   Lawlor A. 2001 June 26. A new kind of life line? Globe and Mail; Sect L:3.

or   11. Lawlor A. A new kind of life line? Globe and Mail 2001 June 26; Sect L:3.

Unsigned newspaper article

APA   Scientists shed light on firefly flickering. (2001, June 29). *The Globe and Mail*, p. S2.

CSE   [Anonymous]. 2001 June 29. Scientists shed light on firefly flickering. Globe and Mail; Sect S:2.

or   12. [Anonymous]. Scientists shed light on firefly flickering. Globe and Mail 2001 June 29; Sect S:2.

Government document

APA   Human Resources Development Canada. (1998). *Changing patterns in women's employment.* Ottawa: Queen's Printer.

CSE   Human Resources Development Canada. 1998. Changing patterns in women's employment. Ottawa: Queen's Printer. 460 p.

or   13. Human Resources Development Canada. Changing patterns in women's employment. Ottawa: Queen's Printer; 1998. 460 p.

Professional website

APA   Health Canada. (2001, November). *Preventing skin cancer.* Retrieved 22 February, 2004, from http://www.hc-sc.gc.ca/english/iyh/diseases/cancer.html

CSE   Preventing skin cancer [Internet]. 2001 Nov. Health Canada; [updated 2004 Feb; cited 2004 Feb 22]. Available from: http://www.hc-sc.gc.ca/english/iyh/diseases/cancer.html

or   14. Preventing skin cancer [Internet]. Health Canada; 2001 Nov [updated 2004 Feb; cited 2006 Feb 22]. Available from: http://www.hc-sc.gc.ca/english/iyh/diseases/cancer.html

Article in an online journal

APA   Pechenik, J.A., Wendt, D.E., & Jarrett, J.N. (1998). Metamorphosis is not a new beginning. *BioScience Online, 48*, 901–910. Retrieved November 17, 1998, from http://www.aibs.org/latitude/latpublications.html

CSE Pechenik JA, Wendt DE, Jarrett JN. 1998. Metamorphosis is not a new beginning. BioSci Online [Internet]. [cited 1998 Nov 17]; 48:901–10. Available from: http://www.aibs.org/latitude/latpublications.html

or 15. Pechenik JA, Wendt DE, Jarrett JN. Metamorphosis is not a new beginning. BioSci Online [Internet] 1998 [cited 1998 Nov 17]; 48:901–10. Available from: http://www.aibs.org/latitude/latpublications.html

CLOSING COMMENTS

As the examples in this chapter show, there is considerable variability in the way that sources are documented, both in the text of a paper or report and in the *References* section. The two systems highlighted here—and remember that these are just two of a vast range of methods—vary widely in places, and to complicate matters further, they are constantly changing, as new ways of conducting research demand updated guidelines for documenting sources.

The examples in this chapter should cover most of the material that you are likely to cite in an undergraduate paper. If you encounter a less common kind of reference, consult the appropriate style guide. Remember that some of the websites mentioned earlier in this chapter may have more up-to-date information on new or revised guidelines. In addition, your college or university department may have its own website with information on how to document sources. It's often a good idea to check here first in case there are particular conventions or a particular style preferred by your department.

REFERENCES

American Psychological Association. (2001). *Publication manual of the American Psychological Association* (5th ed.). Washington, DC: American Psychological Association.

Council of Science Editors. (2006). *Scientific style and format: The CSE manual for authors, editors, and publishers.* (7th ed.). Bethesda, MD: Council of Biology Editors.

chApter 12

Giving a Seminar or Oral Presentation

For some students the prospect of standing in front of a class to give a seminar can be terrifying. The reason some students are terrified of having to speak to a group is almost always that they are afraid of appearing foolish by not knowing what to say or how to answer questions. But there is no reason you can't give a good presentation even if your knees are knocking when you begin—you just have to be prepared. If you think about all the bad seminars you've heard in class, you'll probably find that the reason you couldn't follow what was going on was that the speaker jumped from topic to topic, or missed out crucial segments of an argument, or took for granted things that you didn't know about. The good seminars you have heard were more likely well organized and systematic, leading you through the material being discussed in a logical manner. Some individuals are naturally more comfortable in front of an audience than others, and these students do have a slight advantage. However, you will find that even if public speaking is not one of your natural talents, you can still achieve decent grades by following a few simple rules. The two most important of these are *be prepared* and *be organized*.

MAKING PREPARATIONS

KNOW YOUR TOPIC
For the purposes of an in-class seminar, you are the expert and will probably know more about your topic than any of the other students present. You need to show your audience that your grasp of the subject matter goes beyond what you include in your talk. If you know no more than what you present, you will not be able to answer questions. The more background reading you do, the more you will have to fall back on when someone asks you a question. This can be a confidence booster.

CONSIDER YOUR AUDIENCE
It is a mistake to prepare a seminar based on what you know about your topic. In fact, you should approach the talk from precisely the opposite direction: put yourself in the mind of your audience. If *you* were sitting in class instead of

standing up at the front, what would you expect of the speaker? How much will the typical audience member know about this topic? What will the typical audience member find most interesting? Most relevant? What can you take for granted as common knowledge in the context of the course? If you combine these with your own question—*What do I want my audience to know?*—then you have the basis for setting up your talk.

PLAN YOUR PRESENTATION

Giving a presentation involves much more than writing an essay and then reading it out to the class. Remember that talking to people requires a very different style of presentation. By the time you are asked to give an in-class presentation, it is likely that you will have sat through hundreds of lectures. Think about the ones you enjoyed most and what it was about them that made them interesting. If you do that, you will realize that your best lecturers were the ones who seemed the most prepared, who spoke without reading directly from their notes, who used a range of visual aids, and who seemed animated and interested in what they were talking about. You can be just as interesting by following some of these suggestions:

- **Don't write out everything you plan to say.** You should not write out your whole talk. Unless you are a skilled reader, the presentation will sound awkward and monotonous. Instead, draw up an outline that will serve as a guide as you move through your talk. This will help keep you on track. You should also prepare notes—perhaps on index cards—for each of the points that you are planning to discuss. Because you can't read rough notes to your audience, you will be forced to use your own words and, likely, a more natural speaking style. If you are worried that you may freeze when you first begin, then you could write out the first paragraph of what you want to say, just to get you started.
- **Consider preparing an outline for your audience.** Having a copy of your outline will give your audience something to follow along with as you talk. Typically, you would base this outline on the one you use to organize your talk, but you may want to include additional detail and a bibliography for your audience.
- **Use visual aids.** Having visual aids serves several purposes. First, visual aids can attract and focus the audience's attention. If you are likely to become self-conscious when standing in front of a group, you will be more at ease when all eyes are on your visual aids and not on you. Second, visual aids provide another form of lecture notes to

remind you what you need to say. If you go as far as to use software such as PowerPoint to prepare your presentation, instead of using slides or overhead projections, you will have considerable flexibility to present your information in a variety of ways, for maximum effect.

- **Rehearse your talk.** The better you know what you are going to say by rehearsing your talk in advance, the smoother your presentation will be once you deliver it to your audience. A couple of practice runs will show you where the weak points in your seminar are and will also let you know if you are running over or under time.

GIVING YOUR TALK
DRESS COMFORTABLY

Dressing comfortably means not overdressing but also not dressing down for the occasion. It looks odd if you generally come to class in jeans and T-shirt and you give your presentation in a suit. On the other hand, don't go too far the other way: ripped jeans and an old T-shirt may be too informal for giving a seminar. Something that's clean, comfortable, and casual sets the right tone.

GIVE YOURSELF TIME AT THE BEGINNING

If you have equipment to set up or other preparations to make, be sure to do this before the class begins if you can. If everything is ready, you won't get flustered trying to get your props set up with your classmates looking on.

BEGIN WITH AN OVERVIEW

If the audience knows how the talk will flow, they will be able to understand what you are doing as you move from one point to another. Introduce your topic, then give a brief statement of the main areas that you will discuss. Again, an overhead or handout with the topic outline is useful because the audience can get familiar with it as you work through it.

PROJECT YOUR VOICE

When you speak, be sure you're loud enough that everyone in the classroom can hear you. Also, try to put some feeling into what you say. It is difficult to remain attentive to even the most interesting presentation delivered in a monotone.

DON'T BE APOLOGETIC

The worst way to start a talk is by saying: "You'll have to forgive me, I'm really nervous about this," or "I hope this projector is going to work properly." Even if you are nervous, try to create an air of confidence.

Maintain eye contact with your audience

Look around the room as you speak. When you look at individuals, you involve them in what you are saying. Also, as you scan the faces in front of you, you can monitor for signs of boredom or incomprehension and can adjust your talk accordingly.

Work with your visual aids

If you have visual aids, then take advantage of them; but remember that the visual material should only enhance your talk, not deliver your talk on its own. Some of the following guidelines may help you when you are using visual aids:

- When you are making a point from your overhead, try to use different words and expand on what is there.
- Give your audience enough time to read through each one. Your audience will find it frustrating to see overheads or slides flash by before they've had a chance to take them in.
- Explain your figures. If it's a graph, describe what the x- and y-axes represent, then explain what the graph shows. If it's a diagram, take the audience through it step by step; you may be familiar with the material, but your audience might not be. Later in this chapter we will give you some guidance on how to prepare visual materials to make the strongest impression.

Don't go too fast

A good talk is one that is well paced. If you're discussing background information that everyone is familiar with, you can go over it a little faster; if you're describing something complex or less familiar, go slowly. It often helps to explain a complicated point a couple of times in slightly different ways. Don't be afraid to ask your audience if they understand. Almost certainly, someone will speak up if there is a problem. You should also be careful not to talk too fast. When you're nervous you may find that you are speaking much more quickly than you would in normal conversation. Prepare yourself before you open your mouth; take a few deep breaths then make a concerted effort to speak slowly. In fact you should try to speak *more* slowly than you would in private conversation. It may sound to you that you're crawling along, but to the listener it is likely to seem like a comfortable pace. Pay attention to the way your own instructors deliver their lectures. You'll notice that the ones who are most effective are those who proceed at a measured pace.

MONITOR YOUR TIME ALLOTMENT

As well as pacing your delivery, you should try to ensure that you aren't going to finish too quickly, or worse, go over your allotted time. If you've rehearsed your talk, you should know roughly how long it will take. Remember to take into account that people might ask questions as you are talking. Ideally, you should plan to make your talk a little shorter than the amount of time you have available so that you have some leeway to answer questions.

END STRONGLY

Don't let your talk fade away at the end. You should finish by summarizing the main points you have made and drawing some conclusions. These conclusions should be available on your visual material so that they can be left there for the discussion. If you can raise some questions in your conclusions, this will set you up for the question period to follow.

BE PREPARED FOR QUESTIONS

The question period is a time when you can really make a good impression. This is an opportunity for you to demonstrate your thorough understanding of the topic and even to reinforce one or two points that you think may have been missed. If you know your material well, you should have no problems in dealing with the content of the questions, but the manner in which you answer these questions is important too.

- It's a good idea to repeat a question if you are in a large room where everyone may not have heard it. That should also solidify the question in your mind.
- If you didn't hear or didn't understand a question, don't be afraid to ask the person who asked it to repeat it.
- Keep your answers short and to the point. Rambling answers are not helpful to anyone.
- If you don't know an answer, say so. It's okay to admit that you don't know everything—as long as you don't do this for every question. And, certainly, it's better to admit that you don't know an answer than to guess or to make up a response that everyone will know is not correct.

PREPARING VISUAL AIDS

Only a few years ago, the only ways for a student (or a professor) to illustrate points in a talk were to use photographic slides, which were expensive and inconvenient to prepare, or to make overhead transparencies, which usually had to be hand-written. Today, with the availability of graphic presentation software, as well as laptop computers and video projectors, your ability to use visual aids in a presentation is limited only by your own ingenuity and your instructor's willingness to let you use the technology in class.

For instance, you could develop a PowerPoint presentation that includes video clips and sound as well as animated diagrams. If you have access to this kind of software at home but won't have access to a computer or projector for your talk, you could still prepare your presentation this way and print the slides onto transparencies, which you could display on an overhead projector. Even if you don't have access to PowerPoint or similar software, you could type up the main points of your talk on a word processor, print them out, and photocopy them onto transparency sheets along with any pictures and diagrams you have.

The suggestions that follow apply to any kind of material that you might want to display during a talk.

KEEP IT SIMPLE

This is the cardinal rule and applies to every aspect of your visual aids. It is much better to put too little material on a slide than too much.

- **Use a plain font.** Any word-processing or graphics package will give you lots of font options. Unless you need a fancy one for a specific reason, stick with the simple ones such as **Times New Roman** or **Arial**. You might go as far as using **Comic Sans**, but avoid fonts that are too elaborate. A font like *Lucia Handwriting* or **American Text** can get irritating after a few slides.
- **Choose an appropriate font size.** The last thing you want on your overheads is writing that is too small to be seen. The regular 12-point font you use for your papers will almost certainly be too small when placed on an overhead projector. The minimum size you can use will depend upon how far the projector is from the screen, so you should try making a trial slide with fonts of different sizes; then check them out in the room where you'll be presenting your talk.
- **Use a simple background.** If you're using PowerPoint to make your slides, choose a plain background and use the same one on every slide.
- **Don't overuse colour or other effects.** Unless you have a good reason for doing so, avoid multicoloured slides or other kinds of flashy effects.

- **Don't put too much information on one slide.** If you treat your slides as a script, then you'll be tempted to read directly from them. Instead, make your point briefly on the slide, then expand on the material as you talk. This will make your presentation sound much more natural and professional. If you have a diagram or graph, use the simplest version that you can.
- **Don't use too many slides.** The number of slides you use will depend to some degree on your topic and the kind of material you are presenting, but if you are trying to determine a reasonable number of slides to use during your presentation, a rule of thumb is to have no more than one slide for each minute of your talk. You'll go through some quickly, but you'll need to take more time with others.
- **Use simple transitions.** If you're using PowerPoint, you have the option of introducing text onto a slide in a variety of ways, even letter by letter if you wish. Stay away from such special effects as much as possible. Display everything that relates to a single point on one slide, if possible. It is tiring for the audience to have the information displayed in tiny portions; they don't need to be kept in suspense.

Keep it organized

The second fundamental rule of using visual aids is to make sure your material is well organized. If you use a consistent organizational scheme, the audience will become used to it and will be able to follow along more easily.

- **Begin with a title slide.** A title slide sets the tone and orients the audience to your topic. It should contain the title, your name, and the name of the course.
- **Have an overview slide.** An overview slide should give an outline of your talk so that your audience knows what to expect.
- **Use headings and subheadings.** Most of your slides should be in point form, using numbers or bullets, with headings and subheadings. If you do this, the audience will be able to tell which are your main points and which are elaborations.
- **Consider section breaks.** If your talk falls naturally into several sections, you could start each one with a new title page. Anything that allows the audience to see the structure of your talk is worth doing.
- **End with a summary and/or conclusions slide.** If your topic is a review, you should summarize the main features of your talk at the

end. If you are making an argument, then you should present your conclusions at the end.

- **Keep your overheads in order.** If you are using photographic slides or overhead transparencies, make sure that they are in the correct order—and in the correct orientation—when you start, and be sure to place your transparencies in an ordered pile as you use them. You may have to refer to one later, and you don't want to be shuffling through a disorganized pile in order to find the one you want.

MAKING POSTERS

One form of presentation that has become very popular at scientific conferences as an alternative to an oral presentation is the poster. A poster is a self-contained display that uses a mix of text and graphics to describe a project from start to finish. The author typically stands by the poster for a period of time so that he or she can explain the details of the study and answer questions from other conference participants. This form of presentation is now being used in some senior university courses, particularly fourth-year honours thesis courses.

While this may seem to be a simple option, making an effective poster requires a lot of planning and a little skill. A poster should not be just a condensed version of your paper printed with a slightly larger font. The ideal poster should be eye-catching so that someone walking by will be tempted to stop and look it over; it should be self-contained so that all of the essential aspects of your study can be understood, even if you are not there to explain them; and it should be laid out in a logical fashion so that the reader can follow the sequence easily.

THE PHYSICAL APPEARANCE

When poster sessions were first introduced, some were no more than a printed version of the text of an oral presentation—nothing more than several sheets of typescript containing large amounts of text. Now, those who prepare posters have become much more sophisticated. Of course the availability of graphics software has made preparations much easier, but there is a recognition that posters have their own structure, just like a talk. There are two ways to construct a poster: one is to prepare each of the elements on separate sheets of paper mounted on some form of backing, which are then pinned individually to the poster board. The second, and much more popular, way to create a poster is to prepare it as a single, very large PowerPoint slide. This can be printed using a large format printer and mounted as a single unit. Most commercial printing establishments have the capability to print posters like this. They can even laminate the poster to protect it from tearing. A poster prepared in this way can look very professional.

The typical poster at a scientific conference will be about 2 by 1 metres in size and will contain several separate sections. Some of these will contain limited amounts of text printed in a fairly large font; the rest will contain graphical material, which could be diagrams, photographs, figures containing data, or tables. Each section may be mounted on a sheet of posterboard, or the whole poster may be generated onto a single laminated sheet. We will describe all of these aspects below, but Figure 12.1 gives you an idea of what a typical poster looks like. The following are some guidelines for enhancing the appearance of your poster:

CHOOSE THE RIGHT FONTS

All of the text should be legible from a distance of up to 2 metres. This means that the minimum font size you should use is 18-point. As we suggested for your overheads, use plain fonts. Do not use all capital letters, since they are much more difficult to read than combinations of upper- and lower-case letters. Your title should be even larger so that it can be read from a distance of 5 metres or so.

CHOOSE THE RIGHT MATERIALS

As we mentioned, the easiest way to prepare a poster is to use PowerPoint. You can set the dimensions of individual slides in PowerPoint so that one slide can be as large as 2 metres by 1 metre. You should have been given information about how big the poster presentation board will be, so you know how big you can make your poster, although before beginning you should check with whoever will print the poster to see what their size limits are. Once you have the dimensions, you can insert text and figures, and then rearrange them so that it is easy to follow the sequence of your study. Once you have done that, it is just a matter of sending the file to the printer.

If you don't have the resources to make a full poster like this, you can still make your poster look attractive. You can still use PowerPoint, but this time make the slides regular size and print them on regular paper. After you have done this, you should mount each element of your poster onto a piece of posterboard—heavy cardboard similar to that used for mounting photographs. If this is not available, you can use bristol board, which is thinner. Whatever you do, don't use unmounted pieces of paper.

FIGURE 12.1

Title of a Typical Poster Presentation
A. Student and B. Professor. University of Southern
Ontario, Fort Erie, Ontario, Canada

Introduction

Brief statement of
the problem you will
be investigating

Include hypotheses

Method

Only the essential
details about the
experimental
conditions and the
procedure

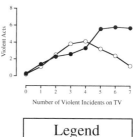

Legend

Apparatus

Only the essential
details. A
diagram or
photograph may
be helpful

Results

Use data figures to
illustrate your main
findings

Summary table if
appropriate

Conclusions

Briefly state the main
findings and your
major conclusions

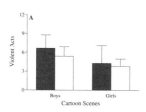

Legend

Legend

References
(Optional)

MAKE AN EFFORT TO ATTRACT PEOPLE'S ATTENTION

Judicious use of colours or pictures (as long as they are relevant) will make your poster stand out from others. But don't go too far by making the poster garish.

THE ESSENTIAL ELEMENTS
TITLE

Include the title of your poster, your name, and the name of your instructor or supervisor. Use upper- and lower-case letters and make them large enough to be legible from a distance of about 5 metres. The title should be mounted by itself on a separate sheet, usually at the top of your poster.

TEXT

The poster should include the same essential elements as a journal article, in roughly the same order: *Introduction* and *Hypotheses*; *Method*; *Results*; and *Conclusions*. You might also include a very short reference list.

GRAPHICS

For posters, it really is true that a picture is worth a thousand words—or at least a couple of hundred. Because you want to keep the text to a minimum, use illustrations, diagrams, and graphs wherever possible, and try to make them self-explanatory.

THE FORMAT
TEXT

- **Be concise.** You should keep the amount of textual material on your poster to a minimum. The reader should be able to read through all of the text in the poster in just a couple of minutes. That means you should not have more than about 400 words of text. Keep asking yourself whether the text you have written is absolutely essential; if not, delete it.

- **Use point form.** Given the space limitations, you can save words by writing brief, declarative sentences, or write in point form.

- **Be clear.** If you have limited space, you won't be able to put in any qualifying statements; save those for when someone asks you about the project. Also, the typical poster reader will look at the title first, look at the statement of the problem next, and then read the conclusions. This will determine whether the poster is worth the additional effort of reading through the methods and the data. Therefore, you should have strong and clear statements in both the first and the last sections.

GRAPHICS

- **Be relevant.** Include only illustrations and figures that relate to the main theme of your study; there is no room for extraneous information.
- **Keep it simple.** Wherever possible, create the simplest figures compatible with clarity. This is to your own benefit, too, because it will save you from having to explain what might be a very complicated diagram.
- **Be logical.** Arrange your figures in a logical sequence so that the reader can follow your thinking with respect to the project.

LAYOUT

- **Plan for the available space.** Before you begin to design your poster, find out the required dimensions; typically it will be about 2 metres long by 1 metre high, but dimension requirements may differ. Once you know this, you can figure out where you will place the various components of the poster. A simple way to do this is to use graph paper (cut proportionally to the dimensions of the posterboard) and small pieces of paper representing your text and graphics to see how you might arrange them. Don't let the space become too cluttered. Try to arrange the papers so that there is space on the top and sides and between each section.
- **Arrange the elements logically.** You want to lead the reader through each element of the poster in right sequence. The easiest route for someone standing in front is to start at the top, left-hand corner and work down, then move over to the right and start again at the top. So, the best way to arrange your poster is in a series of imaginary columns. If it's possible, use the leftmost column for your *Introduction* and *Method* section, the centre columns for your data, and the rightmost column for your *Conclusions* and *References*. If the board dimensions don't allow you to do this, consider using arrows or other pointers to lead the reader through.
- **Label your illustrations.** Although your illustrations should be self-explanatory, a label or a brief description will help to orient the reader.
- **Be neat.** Try to arrange the various elements so that they are balanced,

with equal amounts of space surrounding them. Make small labels to go above each section (or include them on the mounting board containing that section). When you paste your text and graphics onto the mounting board, you should leave a narrow border around the outside to contrast the paper with the background.

- **Do a final run-through.** Before you actually put your poster up, lay it out on the floor to see if everything fits. Use markers to indicate the actual dimensions of the posterboard, then put each piece in its proper place. This is the point when you can fine-tune the arrangement and see if there's anything you can alter to make it easier to read.

CHAPTER 13

WRITING EXAMINATIONS

Before your university career is over, you will have written many examinations. They will come in a variety of forms ranging from true/false and multiple-choice tests to short-answer, problem-solving, and essay exams. There are a number of books available to give you detailed instructions on how to approach different kinds of test. In this chapter we offer you some general suggestions and strategies to deal with some of the most common kinds of examination.

GENERAL GUIDELINES

BEFORE THE EXAM

No matter what type of examination you take, you need to be prepared. This does not mean sitting down a couple of nights beforehand and trying to read through and remember everything in your textbook. It does not even mean reading passively through your notes and texts once a week throughout the term. Studying is an active process, and if you develop good study skills you will be well on your way to success in whatever exams you may take.

The strategy suggested here is the one developed by the study-skills counsellors at the University of Western Ontario. You can adapt it to fit your own needs. The most important thing to remember is to be organized and to use your time effectively. For more information on study skills and exam writing see Fleet, Goodchild, and Zajchowski (2006).

There are six steps to consider in preparing for an exam:

1. getting a perspective
2. learning the material
3. consolidating material and anticipating questions
4. simulating the test
5. filling in the gaps
6. doing a final review

GETTING A PERSPECTIVE

Exam preparation has to begin long before the exam period itself. In fact, you should start organizing and collating your materials as soon as you have the course outline and know what kind of exams you will be taking. Then, as you progress through the term, you should be reading and putting the notes you take in a form that will be useful when the exam finally comes around.

As the exam period approaches, gather together all the materials you have accumulated during the course—textbook, course outline, lecture notes, notes you have made from the textbook, and so on. Skim through this material to remind yourself of the main topic areas. Even at this stage, you should be able to identify general areas from which questions might be drawn.

LEARNING THE MATERIAL

You can make this stage much easier if you have spent some time after each lecture reviewing both your notes and the text. Weekly review will help you remember important material and relate new information to old. If you don't review regularly, at the end of the year you'll be faced with relearning rather than remembering.

As you review, condense and focus the material by writing down in the margin key words or phrases that will trigger whole sets of details in your mind. The trigger might be a word that names or points to an important theory or definition, or it might be a quantitative phrase such as "three factors affecting the development of schizophrenia" or "four classes of operant conditioning."

Sometimes you can create an acronym or a nonsense sentence that will trigger an otherwise hard-to-remember set of facts—something like the mnemonic "Oh, Oh, Oh, To Touch And Feel A Green Vervet At Home" for the initial letters of the 12 cranial nerves. Since the difficulty of memorizing increases with the number of individual items you are trying to remember, any method that will reduce the number of items to be memorized will increase your effectiveness.

At this stage, you can benefit from rewriting your notes and condensing them so that you can go through them quickly during your final review. This is also the time to make sure you understand all the material. Trying to learn by rote something you don't understand is far more difficult than simply hanging facts on a well-understood framework.

Whatever your study plan, don't simply read through your text and other course materials without making notes, asking questions, or solving problems. Something that seems quite straightforward when you read it may turn out to be much less clear when you have to write about it.

CONSOLIDATING AND ANTICIPATING

Here you should be thinking specifically about what questions may be on the exam. The best way to do this is to analyze the course material and then try to make up questions based on it. Rephrasing the material in the form of questions that might be asked should make it easier to recognize and remember when you are in the exam room. If you are able to obtain copies of old exams used previously in the course, these can be useful both for seeing the types of question you might be asked and for checking on the thoroughness of your preparation. If old exams aren't available, you might get together with friends who are taking the same course and ask each other questions. Just remember that the most useful review questions are not the ones that require you to recall facts but those that force you to analyze, integrate, or evaluate information.

SIMULATING THE TEST
Set a hypothetical exam for yourself, based on old exam questions and/or ones that you have made up. Then find a time when you will be free of interruptions and write the exam as if it were the real thing. Although this takes a lot of self-discipline, it's an excellent way to find out your strengths and weaknesses.

FILLING IN THE GAPS
By now you should have a good idea of which areas you know fairly well and which ones need further study. Go over these areas carefully. Don't waste time on things that you know well: just fill in the gaps.

DOING THE FINAL REVIEW
The day before the exam, go over your condensed notes and rehearse some possible questions. At this stage you should have done all of the basic work to make sure you understand and remember the material. What you need now is to get yourself into the best possible frame of mind to write the exam.

TEST ANXIETY
Most students feel nervous before tests and exams. Writing an exam of any kind imposes strong pressures. In an essay exam, because the time is restricted, you can't edit and rewrite the way you can in a regular essay; and because the questions are restricted, you must write on topics you might otherwise choose to avoid. If you are writing a multiple-choice exam, often you don't know whether you are interpreting the questions correctly; and if there is a penalty for guessing, you have the additional stress of deciding whether to mark an answer that you aren't sure is correct.

To do your best on an exam you need to feel calm—but how? It may be of some small comfort to know that a moderate level of anxiety is beneficial: it keeps you alert. It's when you are overconfident or paralyzed with fear that you run into difficulties. There are many strategies for coping with test anxiety, but perhaps the best general advice is to try to control your stress in a positive way. Give yourself lots of time to get to the exam so that you don't need to worry about traffic jams, transit delays, or last-minute room changes. At the exam, focus your attention on your own work rather than concerning yourself with how other students might be performing. Don't keep generating negative "what if" possibilities: if you have studied well, you should be well prepared for any question you may have to answer. Even if you can't turn off your worries, you can reduce them to a point where you will be able to perform well.

AT THE EXAM

If you look at the question paper and your first reaction is *I can't do any of it!* force yourself to keep calm; take several slow, deep breaths to relax, then decide which question you can answer best. Even if the exam seems impossible at first, you can probably find one question that looks manageable: that's the one to begin with. It will get you rolling and increase your confidence. By the time you have finished your first answer, you are likely to find that your mind has worked through the answer for another question.

WRITING AN ESSAY EXAM

READ THE EXAM

An exam is not a hundred-metre dash. Instead of starting to write immediately, take time at the beginning to read through each question and create a plan. A few minutes spent on thinking and organizing will bring better results than the same time spent writing a few more lines.

APPORTION YOUR TIME

Read the instructions carefully to find out how many questions you must answer and to see if you have any choice. Subtract five minutes or so for the initial planning, then divide the time you have left by the number of questions you have to answer. If possible, allow for a little extra time at the end to reread and edit your work. If the instructions on the exam indicate that not all questions are of equal value, apportion your time accordingly.

CHOOSE YOUR QUESTIONS

Decide on the questions that you will do and the order in which you will do them. Your answers don't have to be in the same order as the questions. If you think you have lots of time, it's a good idea to place your best answer first, your poorest answers in the middle, and your second-best answer at the end in order to leave the reader on a high note. If you think you will be rushed, though, it's wiser to work from best to worst; that way you will be sure to get all the marks you can on your good answers, and you won't have to cut a good answer short at the end.

READ EACH QUESTION CAREFULLY

As you turn to each question, read it again carefully and underline all the key words. The wording will probably suggest the number of parts your answer should have; be sure you don't overlook anything—this is a common mistake when people are nervous. Since the verb used in the question is usually a guide for the approach to take in your answer, it's especially important that you interpret the key words in the question correctly. In Chapter 6 we summarized what you should do when you are faced with words like *explain*, *compare*, *discuss*, and so on; it's a good idea to review this list before you go to the exam (*see page 56*).

MAKE NOTES

Before you even begin to organize your answer, jot down key ideas and information related to the topic on rough paper or the unlined pages of your answer book. These notes will save you the worry of forgetting something by the time you begin writing. Next, arrange those parts you want to use into a brief plan, and use numbers to indicate their order (that way, if you change your mind, it will be easy to reorder them). At the end of the exam, you may have to submit any notes you've made with the rest of your paper, so be sure to cross out these notes so that the person marking your paper won't think they are your actual answers.

BE DIRECT

Get to the points quickly and use plenty of examples to illustrate them. In an exam, as opposed to a term paper, it's best to use a direct approach. Don't worry about composing a graceful introduction: simply state the main points that you are going to discuss, then get on with developing them. Remember that your paper will likely be one of many read and marked by someone who has to work quickly; the clearer your answers are, the better they will be received.

For each main point give the kind of specific details that will prove you really know the material. General statements will show you are able to assimilate information, but they need examples to back them up.

WRITE LEGIBLY

Poor handwriting makes readers cranky. When the marker has to struggle to decipher your writing, you may get poorer marks than you deserve. If your writing is not very legible, it's probably better to print. You should also write on every second or third line of your booklet: this will not only make your writing easier to read, but leave you space to make changes and additions if you have time later on.

KEEP TO YOUR TIME PLAN

Keep to your plan and don't skip any questions. Try to write something on each topic. Remember that it's easier to score half marks for a question you don't know much about than it is to score full marks for one you could write pages on. If you find yourself running out of time on an answer and still haven't finished, summarize the remaining points and go on to the next question. Leave a large space between questions so that you can go back and add more if you have time. If you write a new ending, remember to cross out the old one—neatly.

REREAD YOUR ANSWERS

No matter how tired or fed up you are, reread your answers at the end, if there's time. Check especially for clarity of expression; try to get rid of confusing sentences, and improve your transitions so that the logical connection between your ideas is as clear as possible. Revisions that make answers easier to read are always worth the effort.

WRITING A MULTIPLE-CHOICE EXAM

Many students are terrified of multiple-choice exams. They worry that they will need to know every minute detail about the material, or that the questions will be ambiguous. In some cases this is true, but the problem is often related more to the way students approach the test than to their knowledge of the material or to the test itself. The suggestions below are based on the strategy for writing multiple-choice exams proposed by Fleet, Goodchild, and Zajchowski (2006).

1. **Cover up the answers.** This will force you concentrate all of your attention on the question. One of the most common mistakes students make is misreading the question because they read it through too quickly.
2. **Read and process the question.** Take your time. Make sure you understand *exactly* what the question is asking. If necessary, rephrase it in your own words. Pay attention to qualifying words such as "always," "only," or "never."

3. **Predict a possible answer.** Before looking at the alternatives, see if you can answer the question on your own. If you can recall a possible answer, then you're less likely to be fooled by the alternatives that are listed. Although your memory may be jogged by seeing the alternatives, sometimes these are very similar and can be confusing.

4. **Check the format of the question.** Are there combinations of alternatives, such as "all of the above," or "*a* and *c*"? If there are, you need to consider them very carefully. Test-setters sometimes make these combinations the right answer because they know that students with a patchy knowledge of course material will latch on to one fact they know, and that only those with a thorough knowledge of the material will recognize that all of the answers listed are correct.

5. **Process each of the alternatives.** Work your way through each alternative, asking yourself two questions. First, ask if the statement is true, regardless of its relation to the question; sometimes the alternatives are factually incorrect. Second, ask what the relationship of the alternative is to the question; a statement may be true, but not relevant to the question.

6. **Eliminate the wrong answers.** If you have been successful in step 5, you may be able to narrow down the field by excluding the alternatives that cannot be correct.

7. **Identify the correct answer.** At this point you may be in a position to select the correct answer. If not:

8. **Reread the question.** If you're still not certain of the best answer, go back to the question again, now that you have had more time to think about. Sometimes a second reading will make the correct answer much more obvious. Finally:

9. **Guess.** If there is no penalty for guessing, then be sure to select an alternative. If you have eliminated more than one alternative, then even a random choice has good odds of being correct. More important, your guess is likely to be an "educated" one, increasing your chance of success even further. If there is a penalty for guessing—that is, if points are deducted for an incorrect answer—then you need to be more careful with this strategy, but it is usually worth taking the chance.

WRITING AN OPEN-BOOK EXAM

If you think that permission to take your books into the exam room is a guarantee of success, be forewarned: do not fall into the trap of relying too heavily on your reference materials. You may spend so much time flipping through pages and looking things up that you won't have time to write good answers. The result may be worse than if you had been allowed no books at all.

If you want to do well, use your books only to check information and look up specific, hard-to-remember details for a topic you already know a good deal about. For instance, if your subject is biochemistry, you can look up chemical formulae; if it is statistics, you can look up equations; if it is psychology, you can look up specific terms or experimental details. In other words, use the books to make sure your answers are precise and well illustrated. Never use them to replace studying and careful exam preparation.

WRITING A TAKE-HOME EXAM

The benefit of a take-home exam is that you have time to plan your answers and to consult your texts and other sources. The catch is that the amount of time you have to do this is usually less than you would have for an ordinary essay. Don't work yourself into a frenzy trying to respond with a polished research essay for each question. Keep in mind that you were given this assignment to test your overall command of the course: your reader is likely to be less concerned with your specialized research than with evidence that you have understood and assimilated the material.

The guidelines for a take-home exam are similar to those for a regular exam; the only difference is that you don't need to keep such a close eye on the clock:

- Keep your introduction short and get to the point quickly.
- Organize your answer in a straightforward and obvious pattern so that the reader can easily see your main ideas.
- Use many concrete examples to back up your points.
- Where possible, show the range of your knowledge of course material by referring to a variety of sources rather than constantly using the same ones.
- Try to show that you can analyze and evaluate material—that you can do more than simply repeat information.
- If you are asked to acknowledge the sources of any quotations or data you use, be sure to jot them down as you go; you may not have time to do so at the end.

REFERENCE

Fleet, J., Goodchild, F., & Zajchowski, R. (2006). *Learning for success: Effective strategies for students.* (4th ed.). Scarborough, ON: Thomson Nelson.

chapter 14

WRITING WITH STYLE

Writing style is no less important in the sciences than it is in any other discipline. It's true that scientific writing does not tend to go in for fancy words and extravagant images: after all, its main goal is to communicate ideas in a clear and straightforward manner. But any style, from the simplest to the most elaborate, can be effective, depending on the occasion and intent. Writers known for their style are those who have projected something of their own personality into their writing: we can hear a distinctive voice in what they write. In any academic discipline, the most effective style is often one that is clear, concise, and forceful.

SETTING THE TONE

What kind of style is *not* appropriate for scientific writing? While there may be occasions for you to express yourself in a free and casual manner, most of the writing you will be asked to do requires a more formal tone. The following are some signs of writing that may be too informal for academic work.

USE OF SLANG

There are very few occasions when the use of a slang word or phrase is appropriate in a science paper. If you described a rat moving quickly down one arm of a maze as "going like a bat out of hell," you might convey the wrong impression to a reader. Another reason for not using slang expressions is that they are often regional and short-lived: they may mean different things to different groups at different times. (Just think of how widely the meaning of the terms *hot* and *cool* can vary, depending upon the circumstances.)

FREQUENT USE OF CONTRACTIONS

Contractions such as "can't" and "isn't" are not usually suitable for scientific writing, although they may be fine for letters or other informal kinds of writing—for example, this handbook. This is not to say that you should avoid using contractions altogether: even the most serious academic writing can sound stilted or unnatural without any contractions at all. Just be sure that when you use contractions in a college or university report you use them *sparingly*, since excessive use of contractions makes formal writing sound chatty and informal.

EXCESSIVE USE OF FIRST-PERSON PRONOUNS

You should try to keep your work from being "*I*-centred." Most scientific writing has an air of formality about it. This means that when you are describing how you carried out an experiment, for example, it shouldn't read like an e-mail to a friend, with "I" beginning every sentence. On the other hand, you don't want your writing to be so convoluted that it is difficult to follow. In the past, formality was created either by using the passive voice ("the water *was poured* from beaker") or by writing in the third person ("*the experimenter* read the list to the participants"). Although there is nothing technically wrong with either of these strategies, your primary concern should be clarity. Sometimes it makes more sense to use "I" or "we," while at other times a passive construction sounds better. You should judge which seems to flow better in the context of what you are writing.

BE CLEAR

CHOOSE CLEAR WORDS

The key to good writing is using clear words. Two tools that will prove indispensable in this regard are a dictionary and a thesaurus.

A dictionary is a wise investment. A good dictionary will not only help you understand unfamiliar, archaic, or technical words or senses but also help you use these words properly by offering example sentences that show how certain words are typically used. A dictionary can also help you with spelling and with questions of usage: if you are uncertain whether a particular word is too informal for your writing or if you have concerns that a certain word might be offensive, a dictionary will give you this information. You should be aware that Canadian usage and spelling may follow either British or American practices, but usually combine aspects of both; check before you buy a dictionary to be sure that it gives these variants.

A thesaurus lists words that are closely related in meaning. It can help when you want to avoid repeating yourself, or when you are fumbling for a word that's on the tip of your tongue. Be careful, though: make sure you remember the difference between *denotative* and *connotative* meanings. A word's denotation is its primary, or "dictionary," meaning. Its connotations are any associations that it may suggest; they may not be as exact as the denotations, but they are part of the impression the word conveys. If you examine a list of "synonyms" in a thesaurus, you will see that even words with similar meanings can have dramatically different connotations. For example, alongside the word *indifferent* your thesaurus may give the following: *neutral, unconcerned, careless, easy-going, unambitious,* and *half-hearted*. Imagine the different impressions you would create if you chose one or the other of these words to complete this sentence: "Questioned about the project's chance of success, he was _____ in his response." In order to write clearly, you must remember that a reader may react to the suggestive meaning of a word as much as to its "dictionary" meaning.

AVOID JARGON AND USE PLAIN ENGLISH

All academic subjects have their own specialized terminology or *jargon*. It may be unfamiliar to outsiders, but it helps specialists explain things to each other. Precise disciplinary jargon may be suitable for informed audiences, such as your instructor. The trouble is that people sometimes use this sort of special technical language unnecessarily, thinking it will make them seem more knowledgeable. "Never use a short word if a longer, more esoteric one will do," seems to be a general rule for many scientific writers. Too often the result is not clarity but confusion.

The guideline is easy: use specialized terminology only when it's a kind of shorthand that will help you explain something more precisely and efficiently to a knowledgeable audience. If your writing seems stiff or pompous, you may be relying too much on jargon, high-flown phrases, long words, or passive constructions. Although sometimes you must use specialized terms to avoid long and complex explanations, the rest of your paper can be written in simple English. At first glance it may not appear so impressive, but it will certainly be a lot easier to understand.

The following passage appeared in a PhD thesis that was submitted just a few years ago:

> By ameliorating schizophrenic proclivity toward inefficiently deploying their attentional capacity, it is not beyond the realm of possibility that this population could become closer to healthy individuals in terms of cognitive and behavioural functioning.

Roughly translated, this means, "If individuals with schizophrenia could pay more attention, they might be better off." Without lapsing into writing that is too informal, it is possible to produce a clear statement that the reader can evaluate on its own merits just by eliminating the excess words. If plain prose will do just as well, stick to it.

Plain words are almost always more forceful than fancy ones. If you aren't sure what plain English is, think of the way you talk to your friends (apart from swearing and slang). Many of our most common words—the ones that sound most natural and direct—are short. A good number of them are also among the oldest words in the English language. By contrast, most of the words that English has derived from other languages are longer and more complicated; even those that have been used for centuries can sound artificial. For this reason you should beware of words loaded with prefixes (*pre-, post-, anti-, pro-, sub-, maxi-,* etc.) and suffixes (*-ate, -ize, -tion,* etc.). These Latinate attachments can make individual words more precise and efficient, but putting a lot of them together will make your writing seem dense and hard to understand. In many cases you can substitute a plain word for a fancy one:

Fancy	*Plain*
accomplish	do
cognizant	aware
commence	begin, start
conclusion	end
determinant	cause
fabricate	build
finalize	finish, complete
firstly	first
infuriate	anger
maximization	increase
modification	change
numerous	many
obviate	prevent
prioritize	rank
proceed	go
remuneration	pay
requisite	needed
sanitize	clean
subsequently	later
systematize	order
terminate	end
transpire	happen
utilize, utilization	use

Suggesting that you write in plain English does not mean that you should never pick an unfamiliar, long, or foreign word; sometimes these words are the only ones that will convey precisely what you mean. Inserting an unusual expression into a passage of plain writing can also be an effective means of catching the reader's attention—as long as you don't do it too often. And, of course, writing clearly does not mean that you should avoid all specialized scientific terms. Just remember that when you use technical language, your instructors will not be impressed by the mere presence of these words: appropriate disciplinary terminology must be used correctly.

BE PRECISE

Always be as precise as you can. Avoid all-purpose adjectives like *major, significant*, and *important* and vague verbs such as *involved, entail*, and *exist* when you can be more specific:

orig.   Catalysts <u>are involved</u> in many biochemical reactions.

rev.   Catalysts <u>speed up</u> many biochemical reactions.

Here's another example:

orig.   The discovery of genetic engineering techniques was a <u>significant</u> contribution to biological science.

rev.   The discovery of genetic engineering techniques was a <u>dangerous</u> contribution to biological science.

(or)

rev.   The discovery of genetic engineering techniques was a <u>beneficial</u> contribution to biological science.

AVOID UNNECESSARY QUALIFIERS

Qualifiers such as *very, rather*, and *extremely* are overused. Saying that something is *very elegant* may have less impact than saying simply that it is *elegant*. For example, compare these sentences:

They devised an <u>extremely elegant</u> hypothesis to explain their data.

They devised an <u>elegant</u> hypothesis to explain their data.

Which has a greater impact? When you think that an adjective needs qualifying—and sometimes it will—first see if it's possible to change either the adjective or the phrasing. Instead of writing

Multinational Drugs made a <u>very big</u> profit last year,

write a precise statement:

Multinational Drugs made an <u>unprecedented</u> profit last year,

or (if you aren't sure whether or not the profit actually set a record)

Multinational Drugs' profit rose 40 per cent last year.

In some cases, qualifiers not only weaken your writing but are redundant because the adjectives themselves are absolutes. To say that something is *very unique* makes as little sense as saying that someone is *slightly pregnant* or *extremely dead*.

CREATING CLEAR PARAGRAPHS

Paragraphs come in so many sizes and patterns that no single formula could possibly cover them all. The two basic principles to remember are these:

1. A paragraph is a means of developing and framing an idea or impression.
2. The divisions between paragraphs aren't random but indicate a shift in focus.

With these principles in mind, you should aim to include three elements in each paragraph:

1. the topic sentence, to indicate to the reader the subject of the paragraph;
2. a supporting sentence or sentences, to convey evidence or develop the argument;
3. a conclusion, to indicate to the reader that the paragraph is complete.

Keep these points in mind as you write. The following sections offer additional advice on creating clear paragraphs.

DEVELOP YOUR IDEAS

You are not likely to sit down and consciously ask yourself, "How will I construct this paragraph?" What comes first is the idea you intend to develop: the structure of the paragraph should flow from the idea itself and the way you want to discuss or expand it.

You may take one or several paragraphs to develop an idea fully. For a definition alone you could write one paragraph or ten, depending on the complexity of the subject and the nature of the assignment. Just remember that ideas need development, and that each new paragraph signals a change in idea.

CONSIDER THE TOPIC SENTENCE

Skilled skim-readers know that they can get the general drift of a book simply by reading the first sentence of each paragraph. The reason is that most paragraphs begin by telling the reader what the paragraph is about, stating the idea to be developed or the point to be made.

Like the thesis statement for the paper as a whole, the topic sentence is not obligatory: in some paragraphs the controlling idea is not stated until the middle or even the end, and in others it is not stated at all but merely implied. Nevertheless, it's a good idea to think out a topic sentence for every paragraph. That way you'll be sure that each one has a definite point and is clearly connected to what comes before and after. When revising your initial draft, check to see that each paragraph is held together by a topic sentence, in which the central idea of the paragraph is either stated or implied. If you find that you can't formulate one, it could be that you are uncertain about the point you are trying to make; in this case, it may be best to rework the whole paragraph.

MAINTAIN FOCUS

A clear paragraph should contain only those details that are in some way related to the central idea. It must also be structured so that the details are easily *seen* to be related. One way of showing these relations is to keep the same grammatical subject in most of the sentences that make up the paragraph. When the grammatical subject is shifting all the time, a paragraph loses focus, as in the following example (Cluett & Ahlborn, 1965):

> orig.   Students at our school play a variety of sports these days. In the fall, football still attracts the most, although an increasing number now play soccer. For some, basketball is the favourite when ball season is over, but you will find that swimming and gymnastics are also popular. Cold winter temperatures may allow the school to have an outdoor rink, and then hockey becomes a source of enjoyment for many. In spring though, the rinks begin melting, and so there is less opportunity to play. Then some students take up soccer again, while track and field also attracts many participants.

Here the grammatical subject (underlined) changes from sentence to sentence. Notice how much stronger the focus becomes when all the sentences have the same grammatical subject—either the same noun, a synonym, or a related pronoun:

> rev.   Students in school play a variety of sports these days. In the fall, most still choose football, although an increasing number now play soccer. When the ball season is over, some turn to basketball; others prefer swimming or gymnastics. If cold winter temperatures permit an outdoor

rink, many <u>students</u> enjoy hockey. Once the ice begins to melt in spring, though, <u>they</u> can play less often. Then <u>some</u> take up soccer again, while <u>others</u> choose track and field.

Naturally it's not always possible to retain the same grammatical subject throughout a paragraph. If you were comparing the athletic pursuits of boys and girls, for example, you would have to switch back and forth between boys and girls as your grammatical subject. In the same way, you will have to shift subjects when you are discussing examples of an idea or exceptions to it.

AVOID MONOTONY

If most or all of the sentences in your paragraph have the same grammatical subject, how do you avoid boring your reader? There are two easy ways:

1. **Use stand-in words.** Pronouns, either personal (*I, we, you, he, she, it, they*) or demonstrative (*this, that, those*), can stand in for the subject, as can synonyms (words or phrases that mean the same thing). The revised paragraph on school sports, for example, uses the pronouns *some, most,* and *they* as substitutes for *students.* Most well-written paragraphs have a liberal sprinkling of these stand-in words.
2. **"Bury" the subject by putting something in front of it.** When the subject is placed in the middle of the sentence rather than at the beginning, it's less obvious to the reader. If you take another look at the revised paragraph, you'll see that in several sentences there is a word or phrase in front of the subject. Even a single word, such as *first, then, lately,* or *moreover,* will do the trick. (Incidentally, this is a useful technique to remember when you are writing a letter of application and want to avoid starting every sentence with *I.*)

LINK YOUR IDEAS

To create coherent paragraphs, you need to link your ideas clearly. Linking words are those connectors—conjunctions and conjunctive adverbs—that show the relations between one sentence, or part of a sentence, and another; they're also known as transition words because they bridge the transition from one thought to another. Make a habit of using linking words when you shift from one grammatical subject or idea to another, whether the shift occurs within a single paragraph or as you move from one paragraph to the next. The following are some of the most common connectors and the logical relations they indicate:

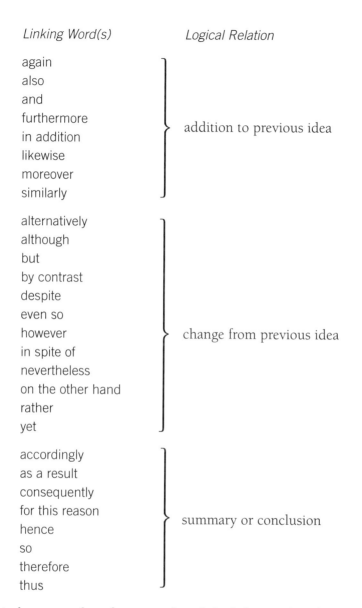

Linking Word(s)	Logical Relation
again also and furthermore in addition likewise moreover similarly	addition to previous idea
alternatively although but by contrast despite even so however in spite of nevertheless on the other hand rather yet	change from previous idea
accordingly as a result consequently for this reason hence so therefore thus	summary or conclusion

Numerical terms such as *first, second*, and *third* also work well as links.

VARY THE LENGTH, BUT AVOID EXTREMES

Ideally, academic writing will have a balance of long and short paragraphs. Avoid the extremes—especially the one-sentence paragraph, which can only state an idea without explaining or developing it. A series of very short paragraphs is usually a sign that you have not developed your ideas in enough detail, or that you have started new paragraphs unnecessarily. On the other hand, a succession of long paragraphs can be tiring and difficult to read. In deciding when to start a new paragraph, remember always to consider what is clearest and most helpful for the reader.

BE CONCISE

At one time or another, you will probably be tempted to pad your writing. Whatever the reason—because you need to write two or three thousand words and have only enough to say for one thousand, or just because you believe that "length is strength" and hope to get a better mark for the extra words—padding is a mistake. You may fool some of the people some of the time, but not often.

Strong writing is always concise. It leaves out anything that does not serve some communicative or stylistic purpose in order to say as much as possible in as few words as possible. Concise writing will help you do better on both your essays and your exams.

AVOID AMBIGUOUS PRONOUNS

You will confuse the reader if you use a pronoun like *it* or *this* without being clear about which nouns they belong to:

> As the light intensity was lowered, the mixture became viscous. It also turned blue.

Does *it* in this example refer to the colour of the light or the colour of the mixture? Repeat the relevant noun if there is any possibility of confusion.

AVOID USING TOO MANY ABSTRACT NOUNS

Whenever possible, choose a verb rather than an abstract noun:

orig. The <u>combination</u> and <u>modification</u> of the lists resulted in <u>confusion</u> between the items.

rev. When the lists were <u>combined</u> and <u>modified</u>, the participants became <u>confused</u> about the items.

AVOID VAGUE QUALIFIERS

Scientific writing should be precise. In particular, you should avoid words such as *quite, very, fairly, some, many,* or *roughly* if you can use a more exact term. You should also avoid the scatter-gun approach to adverbs and adjectives: don't use combinations of modifiers unless you are sure they clarify your meaning. One well-chosen word is always better than a series of synonyms:

orig.   As well as being <u>costly</u> and <u>financially extravagant</u>, the project is <u>reckless</u> and <u>foolhardy</u>.

rev.   The venture is <u>costly</u> as well as <u>foolhardy</u>.

AVOID NOUN CLUSTERS

A recent trend in some writing is to use nouns as adjectives (as in the phrase "*noun* cluster"). This device can be effective occasionally, but frequent use can produce a monstrous pile-up of nouns. Breaking up noun clusters may not always result in fewer words, but it will make your writing easier to read:

orig.   personal digital assistant utilization manual

rev.   manual for using a personal digital assistant

orig.   pollution investigation committee

rev.   committee to investigate pollution

Certain nouns that are used routinely as adjectives—"*reaction* time," "*block* design," "*water* bottle," and so on—are quite acceptable. Again, use your judgment to decide what sounds best.

AVOID CHAINS OF RELATIVE CLAUSES

Sentences full of clauses beginning *which, that*, or *who* are usually wordier than necessary. Try reducing some of these clauses to phrases or single words:

orig.   The solutions <u>that</u> were discussed last night have a practical benefit, <u>which</u> is easily grasped by people <u>who</u> have no technical training.

rev.   The solutions discussed last night have a practical benefit, easily grasped by non-technical people.

TRY REDUCING CLAUSES TO PHRASES OR WORDS

Independent clauses can often be reduced by subordination. Here are a few examples:

orig.   The report was written in a clear and concise manner, and it was widely read.

rev.   Written in a clear and concise manner, the report was widely read.

rev.   Clear and concise, the report was widely read.

orig.   His plan was of a radical nature and was a source of embarrassment to his employer.

rev.   His radical plan embarrassed his employer.

ELIMINATE HACKNEYED EXPRESSIONS AND CIRCUMLOCUTIONS

Trite or roundabout phrases may flow from your pen without a thought, but they make for stale prose. Unnecessary words are deadwood; be prepared to hunt and chop ruthlessly to keep your writing vital:

Wordy	Revised
due to the fact that	because
at this point in time	now
consensus of opinion	consensus
in the near future	soon
when all is said and done	[omit]
in the eventuality that	if
in all likelihood	likely

AVOID "IT IS" AND "THERE IS" BEGINNINGS

Although it may not always be possible, try to avoid beginning your sentences with "It is . . ." or "There is (are) . . .". Your sentences will be crisper and more concise:

orig.   There are only a few days in the year when the phenomenon may be observed.

rev.   The phenomenon may be observed on only a few days of the year.

orig.   It is certain that pollution will increase.

rev.   Pollution will certainly increase.

BE FORCEFUL

Developing a forceful, vigorous style simply means learning some common tricks of the trade and practising them until they become habit.

CHOOSE ACTIVE OVER PASSIVE VERBS

As we have mentioned earlier (Chapter 7), scientists who publish in professional journals are divided on whether you should use the active or passive voice. Traditionally, most have preferred the passive voice ("The rat was placed in the starting box of the maze," rather than "I placed the rat in the starting box") because they believe that its impersonal quality helps to maintain the detached, impartial tone appropriate for a scientific report. Today, more and more scientists are using and recommending use of the active voice because it is clearer and less likely to lead to awkward sentences. Passive constructions tend to produce awkward, convoluted phrasing:

orig.   It <u>has been decided</u> that the utilization of small rivers in the province for purposes of generating hydro-electric power <u>should be studied</u> by our department, and that a report to the deputy <u>should be made</u> by our director as soon as possible.

The passive verbs in this example make it hard to tell who is doing what. This passage is much clearer without the passive verbs:

rev.   The Minister of Natural Resources <u>has decided</u> that our department <u>should study</u> the use of small rivers in the province to generate hydro-electric power, and that our director <u>should make</u> a report to the deputy as soon as possible.

An active verb also creates more energy than a passive one does:

Passive   The data <u>were collected and analyzed</u> by her.

Active   She <u>collected and analyzed</u> the data.

Passive verbs are appropriate in four specific cases:

1. When the subject of the sentence is the passive recipient of some action:

   The politician <u>was heckled</u> by the angry crowd.

2. When you want to emphasize the object rather than the person acting:

   The anti-pollution devices in all three plants <u>will be improved</u>.

3. When you want to avoid an awkward shift from one subject to another in a sentence or paragraph:

   The guppies had adjusted to their new tank but they <u>were eaten</u> by the larger fish a short time later.

4. When you want to avoid placing responsibility or blame:

   The plans <u>were delayed</u> when the builder became ill.

USE PERSONAL SUBJECTS

Most of us find it more interesting to learn about people than about things. Wherever possible, therefore, make the subjects of your sentences personal. This trick goes hand in hand with use of active verbs. Almost any sentence becomes livelier with active verbs and a personal subject:

orig.   The materialistic <u>implications</u> of Darwin's theory led to a long delay before it <u>was published</u>.

rev.   <u>Darwin</u> <u>delayed</u> publication of his theory for a long time because of its materialistic implications.

Here's another example:

orig.   <u>It may be concluded</u> that the reaction <u>had been permitted</u> to continue until it <u>was completed</u> because <u>there was no sign</u> of any precipitate when the flask <u>was examined</u>.

rev.   <u>We may conclude</u> that the reaction <u>had ended</u> because <u>we could not see</u> any precipitate when <u>we examined</u> the flask.

USE CONCRETE DETAILS

Concrete details are easier to understand—and to remember—than abstract theories. Whenever you are discussing abstract concepts, therefore, always provide specific examples and illustrations; if you have a choice between a concrete word and an abstract one, choose the concrete. Consider this sentence:

> Watson and Crick were the first to demonstrate the three-dimensional structure of DNA.

Now see how a few specific details can bring the facts to life:

> Watson and Crick showed that the DNA molecule was arranged as a double helix and that this structure helped explain how genetic material could be replicated.

Suggesting that you add concrete details doesn't mean getting rid of all abstractions; it's simply a reminder to balance them with accurate details. The example above is one occasion when added wording, if it is concrete and correct, can improve your writing.

MAKE IMPORTANT IDEAS STAND OUT

Experienced writers know how to manipulate sentences in order to emphasize certain points. The following are some of their techniques.

PLACE KEY WORDS IN STRATEGIC POSITIONS

The positions of emphasis in a sentence are the beginning and, above all, the end. If you want to bring your point home with force, don't put the key words in the middle of the sentence—save them for last:

orig.   People are less afraid of losing wealth than of losing face in this image-conscious society.

rev.   In this image-conscious society, people are less afraid of losing wealth than of losing face.

SUBORDINATE MINOR IDEAS

Small children connect incidents with a string of *ands*, as if everything were of equal importance:

> Our bus was delayed, and we were late for school, and we missed our class.

As they grow up, however, they learn to *subordinate*—that is, to make one part of a sentence less important in order to emphasize another point:

> Because the bus was delayed, we were late and missed our class.

Major ideas stand out more and connections become clearer when minor ideas are subordinated:

orig.   Spring arrived and we had nothing to do.

rev.   When spring arrived, we had nothing to do.

Make your most important idea the subject of the main clause, and try to put it at the end, where it will be most emphatic:

orig.   I was relieved when I saw my marks.

rev.   When I saw my marks, I was relieved.

VARY SENTENCE STRUCTURE

As with anything else, variety adds spice to writing. One way of adding variety, which will also make an important idea stand out, is to use a periodic rather than a simple sentence structure.

Most sentences follow the simple pattern of subject – verb – object (plus modifiers):

> The <u>dog</u> <u>bit</u> the <u>man</u> on the ankle.
> S   V   O

A *simple sentence* such as this gives the main idea at the beginning and thus creates little tension. A *periodic sentence*, on the other hand, does not give the main clause until the end, following one or more subordinate clauses:

> Because there was little demand for his course, in the following year <u>it was</u>
> <u>cancelled</u>.   **S V**

The longer the periodic sentence is, the greater the suspense and the more emphatic the final part. Since this high-tension structure is more difficult to read than the simple sentence, your readers would be exhausted if you used it too often. Save it for those times when you want to make a very strong point.

VARY SENTENCE LENGTH
A short sentence can add punch to an important point, especially when it comes as a surprise. This technique can be particularly useful for conclusions. Don't overdo it, though—a string of long sentences may be monotonous, but a string of short ones can make your writing sound like a children's book.

USE CONTRAST
Just as a jeweller will highlight a diamond by displaying it against dark velvet, so you can highlight an idea by placing it against a contrasting background:

> orig.   Most employees in industry do not have indexed pensions.

> rev.   <u>Unlike civil servants,</u> most employees in industry do not have indexed pensions.

Using parallel phrasing will increase the effect of the contrast:

> Although she often gave informal talks and seminars, she seldom gave formal presentations at conferences.

USE A WELL-PLACED ADVERB OR CORRELATIVE CONSTRUCTION
Adding an adverb or two can sometimes help you to dramatize a concept:

> orig.   The suggestion is good, but I doubt it will succeed.

> rev.   The suggestion is good <u>theoretically,</u> but I doubt it will succeed <u>practically</u>.

Correlatives such as *both . . . and* or *not only . . . but also* can be used to emphasize combinations as well:

> orig.   Professor Muttucumaru was a good lecturer and a good friend.

> rev.   Professor Muttucumaru was <u>both</u> a good lecturer <u>and</u> a good friend.

> (or)

rev.   Professor Muttucumaru was <u>not only</u> a good lecturer <u>but also</u> a good friend.

Use your ears

Your ears are probably your best critics: make good use of them. Before producing a final copy of any piece of writing, read it out loud in a clear voice. The difference between cumbersome and fluent passages will be unmistakable.

A NOTE ON THE USE OF NEUTRAL LANGUAGE

The primary goals of scientific writing are *objectivity*—a term that implies a lack of bias—and *accuracy*, which means freedom from ambiguity. Although we tend to think of objectivity and accuracy in relation to the presentation and interpretation of data, they also have broader applicability.

In years past, relatively little attention was paid to the potential implications of using certain words. So, for example, referring to someone with a physical impairment as a "cripple" was not considered out of the ordinary. More recently, of course, there has been an increasing sensitivity to language that is not inclusive or that may cause offence. This is not the place to debate whether the pendulum of sensitivity has swung too far in one direction. However, you do need to think about what the words you use might mean to another person.

Let's consider the word "cripple." Suppose a student remarks that a particular professor was such a good instructor that she was able to help even the "math cripples" in the class. In this context, the term is not intended to show disrespect for individuals with physical disabilities; it is a metaphor to highlight the students' lack of math skills. However, it is possible that a person with a physical disability might take offence on hearing that remark. Do you feel that the student should not have used this term? What if the student was talking instead about her finances and remarked that by the time she graduated she would have a "crippling debt load." Should such a comment also be avoided because it contains a term that is potentially offensive to someone?

The point of these examples is not to tell you whether or not you should use particular words, or how far you should go to avoid giving offence. There is no correct answer to the questions above, and there have been long and strenuous debates about "political correctness." Rather than getting involved in such discussions, though, we would suggest that the most important thing is to think about the implications of what you say before you say it.

There are three areas that have attracted particular attention with respect to neutrality: the use of terms referring to men and women; the use of terms referring to different ethnic groups; and the use of terms referring to individuals with disabilities. We will discuss each of these in turn.

TERMS REFERRING TO MALES AND FEMALES

Until quite recently, "man" was widely used to refer to humans in general. It was not considered odd or inappropriate for Charles Darwin to write about *The Descent of Man*, and books with titles like *Man and animal: studies in behaviour* or *Man and beast: comparative social behavior* were quite common. A quick search of the library catalogue at the University of Western Ontario produced 462 items beginning with "*Man and . . .*". In almost all of these cases the word "man" could be replaced by "humans" with no alteration in meaning. In contrast, there were only 76 listings for "*Woman and . . .*", and most of these were referring to "woman" as a female. Curiously, a search on "Men and . . ." and "Women and . . ." produced 230, and 1,247 titles respectively.

Rather than asking whether this use of words is "sexist," as some people argue, ask yourself if it meets the criteria of objectivity and accuracy. Clearly, the use of gender-specific terms in this fashion does not meet the criterion of accuracy, and may not meet the objectivity standard either. If you want to refer to the human species in general, then use "human," "humankind," or "people."

The same guideline can be applied to job descriptions that traditionally were dominated by members of one sex—*fireman*, *policeman*, and *stewardess*, for example. In such cases, a slight modification allows the noun to cover both males and females: *fire fighter*, *police officer*, and *flight attendant* are all perfectly acceptable alternatives.

You should be careful, though, not to carry these modifications to extremes. There are many words that contain the letters "m-a-n" that have a completely different etymology and do not refer to maleness. There is no need to find alternatives for such words as "manufacture," "manage," or "manipulate." You should also recognize that certain words that may originally have had a male connotation have become so fully integrated into the language that to change them would result in a fairly convoluted or unintelligible rewording. "Manhole cover," for example, has a specific referent and no longer implies that only men would go through the hole when it is uncovered. Although some people would argue that even these terms are sexist, the clarity achieved by using such a word seems more important than its etymology.

Another area of difficulty is the use of masculine pronouns to refer to someone who might be male or female. It was once conventional to use "he" or "his" whenever one was talking in general about an individual. So, each of the participants in an experiment was usually referred to as "he" even if some were female, as in "Each of the participants had to complete his written questionnaire before he could be interviewed." Today, using "he" or "his" to refer to a person of either sex is not widely accepted.

You can get around this problem in a couple of ways. If you are writing a paper in which there are many references to individuals who may be male or female, then you might decide to use either the masculine or feminine pronouns exclusively; just be sure to acknowledge when you first use them that they will apply to both. This solution is not ideal, however: even with your acknowledgement that "he" and "his" will be used to refer to individuals of either sex, using the male pronouns hardly seems accurate, particularly if the majority of your participants are female.

If there will only be a few references of this type, then you can say "his or her" and "he or she," as in "Each of the participants had to complete a written questionnaire before he or she could be interviewed." This solution is quite acceptable; however, compound pronouns are cumbersome, so if your report contains more than a few references of this sort, this solution might not be suitable.

The best solution may be to change your singular subject to a plural one, which would allow you to use gender-neutral pronouns *they*, *them*, and *their*: "The participants had to complete *their* written questionnaires before *they* could be interviewed." We would advise against the use of "s/he," which looks odd.

Finally, it is worth saying something about the use of the terms "sex" and "gender." "Sex" refers to the biological distinction between males and females, and any discussion that deals with that distinction should refer to "sex." For example, if it was not important whether you were testing males or females, you would say that, "the sex of the participants was not relevant to this study." "Gender" refers either to a grammatical classification of words or to the cultural, social, or psychological dimensions of maleness or femaleness. Try to use these two terms appropriately.

TERMS REFERRING TO DIFFERENT ETHNIC GROUPS

Historically, Caucasian writers would refer to different ethnic groups using terms developed within the white community. Thus, terms like "Indian," "Eskimo," "Oriental," "Coloured," or "Negro" were used routinely, and often with a pejorative connotation. Little consideration was given to using the terms used by members of these communities themselves.

Beginning in the 1960s with the rise of the civil rights movement in the United States, there was a concerted effort by different ethnic groups to promote descriptors that were accepted within their own communities. As a result, terms like "Afro-American" and "African American" have become the descriptors of choice. Similarly, the term "Eskimo" has been replaced by "Inuit," "Oriental" by reference to specific country of origin, and so on. You should be aware, though, that language is dynamic, so the preferred terms may change over time or vary depending on the context. A good example of this is the term "Indian" used to refer to Native Americans and Canadians. It is still used in some official capacities: for instance, the U.S. government still has a "Department of Indian Affairs," and some Canadian legislation recognizes "status Indians" and "non-status Indians," those who, respectively, are and are not members of bands that have signed treaties with the government. In Canada, the term "Indian" is useful when distinguishing among the three groups of Aboriginal peoples: Indians, Inuit, and Métis. However, among the people themselves, the term "First Nations citizen" is generally preferred, and many individual groups favour and have begun to promote the use of more specific references—Anishnabe, Chippewa, Cree, Nisga'a, Oneida, and so on.

The rule you should follow if you are referring to a particular ethnic group is to find out what name is used within the community you are discussing and use that one.

TERMS REFERRING TO IMPAIRMENT, DISABILITY, AND HANDICAP

As was the case with ethnic group references, there used to be little recognition of people with disabilities as individuals. Instead there was a tendency to use broad terms that perpetuated negative stereotypes: "deaf and dumb," "cretin," and "Mongoloid" were all terms used indiscriminately in both formal and informal writing. With respect to the criteria we set out above, these terms are neither objective nor accurate. Even the words "disability" and "handicap" are sometimes used inappropriately.

A number of years ago, the World Health Organization (1980) proposed a set of definitions to describe conditions that might limit a person's abilities. When you read these, you will see that it makes no sense to put a label on a person because of his or her condition. The conditions proposed by WHO are as follows:

- **Impairment.** *Impairment* refers to the nature of the condition itself. Thus, a person may be blind, or have a profound hearing impairment, or be unable to walk because of a spinal cord injury. These are the impairments.
- **Disability.** An impairment becomes a *disability* when it prevents a person from engaging in his or her normal, day-to-day activities. What

this means is that if there are two people with identical impairments, one may be considered to have a disability and the other may not. For example, a person with limited eyesight may successfully maintain a completely independent lifestyle. That person should not be considered to have a disability. In contrast, a person with exactly the same limitation of vision may find the condition overwhelming and may not be able to cope without considerable outside help. In this case, the vision impairment would be considered a disability.

- **Handicap.** An impairment or a disability becomes a handicap when there are social consequences that are imposed from outside. So, if a person who uses a wheelchair cannot gain admittance into buildings routinely or is denied a job because of his or her condition, that person would be considered to have a handicap.

The reason for giving you these definitions is to help you realize that you must not categorize individuals purely on the basis of their clinically defined conditions. If you keep this in mind, then it becomes easy to see why it makes sense to use a phrase like "people with disabilities" instead of "the disabled," "a person with diabetes" instead of "a diabetic," or "profoundly hearing-impaired" instead of "deaf and dumb." You must remember to emphasize the person and not the condition.

Recently the World Health Organization (2001) revised their classification system to provide a more comprehensive description of a person's ability to function, emphasizing both positive and negative aspects. This system discusses "functioning and disability" as one aspect of the classification and sets this in the context of environmental and personal factors that can influence a person's activities and participation in a wide variety of different areas.

One final point: if you are working with any group of participants from a clinical population or with individuals who have an impairment or disability, do not use the term "normals" to distinguish your non-clinical group from the others; it is demeaning.

CONCLUSION

The question of what constitutes biased writing is a difficult one because there are so many different points of view. No matter how hard you try, it's unlikely that you will be able to satisfy everyone. The most successful strategy when you are writing is to make sure that you do not take traditional assumptions for granted. For example, it is no longer accurate to assume that doctors or researchers are all male, or that nurses and research assistants are female.

Wherever possible you should use specific, or neutral, or inclusive terms, and you should always be alert to the possibility that you may be misrepresenting a group. However, in our view, you should not tie your writing in knots to satisfy all demands for neutrality and correctness. You will have to use your judgment to determine what is acceptable and what is not. You should also be aware that what is acceptable now may change over time, so that a word that is considered appropriate today may not be five years from now.

REFERENCES

Cluett, R., & Ahlborn, L. (1965). *Effective English prose*. New York: L.W. Singer.

World Health Organization. (1980). *ICDH: International classification of impairments, disabilities, and handicaps: a manual of classification relating to the consequences of disease*. Geneva: World Health Organization.

World Health Organization. (2001). *ICDH-2: International classification of functioning, disability and health (Final draft)*. [Web page]. Retrieved August 8, 2001, from the World Wide Web: http://www.who.int/icidh/.

chapter 15

WRITING A RESUMÉ OR A LETTER OF APPLICATION

Whether you are looking for a summer job, applying to graduate school, or seeking permanent employment, eventually you will have to write a resumé and letter of application. The person who reads your application will not have time to read reams of material, so you will need to be brief yet precise.

PREPARING A STANDARD RESUMÉ

Think of a resumé as more than just a summary of facts: think of it as a selling tool, designed to fit an individual employer's or organization's needs. You will need to supply some basic information, but how you organize it and which details you emphasize are up to you. One good strategy is to put your most important qualifications first so that they are noticed at first glance. For most students this means leading with educational qualifications, but for others it may mean starting with work experience. Within each section of your resumé, use reverse chronological order so that the most recent item is at the beginning.

Whatever arrangement you choose, your goal is to keep the resumé as concise as possible while including all the specific information that will help you "sell" yourself. A reader will lose interest in a resumé that goes on and on, mixing pertinent details with trivial ones. On the other hand, experience or skills that may seem irrelevant might actually demonstrate an important attribute or qualification. For example, working as a part-time short-order cook may be significant if you state that this was how you paid your way through university.

Here is a list of common resumé information, along with some suggestions on how to present it:

- **Name.** Your name is usually placed in capital letters and centred at the top of the page.
- **Address and telephone number.** These can come at the top or bottom of the page. If you have a temporary student address, remember to state where you can be reached at other times.

- **Career objectives (optional).** It's often helpful to let the employer know your career goal, or at least your current aim for employment.
- **Education.** Include any degrees or related diplomas or certificates along with the institution that granted them and the dates they were granted. If it will help your case, and if you are short of other qualifications, you may also list courses you have taken that are relevant to the job.
- **Awards or honours.** These may be in a separate section or included with your education information.
- **Work experience.** Give the name and location of your employer, along with your job title and the dates of employment. Instead of outlining your duties (which an employee may or may not carry out well), list your accomplishments on the job, using point form and action verbs. For example:
  - Designed and administered a public awareness survey.
  - Supervised a 3-member field crew.
- **Specialized skills.** This is a chance to list information that may give you an advantage in a competitive market, such as experience with computers or knowledge of a second language. If you have worked as a research assistant, be sure to state the type of work you did and for whom you worked. For example:
  - Assisted Professor William James in laboratory research project on social learning in gerbils, Lakehead University, Summer 2000.
- **Other interests (optional).** Depending on the amount of information you have already included and on the type of employer, you may choose to omit this section. Sometimes, however, including a few achievements or interests, such as athletic or musical accomplishment, will show that you are well-rounded or specially disciplined. Avoid making a long list of items that merely show passive or minimal involvement.
- **References (optional).** If asked to supply the names of people who can give references, be sure to give the complete title, address, phone number, and e-mail address of each one. If you are not asked for references, you don't need to supply names. Increasingly, employers are waiting until they are serious about hiring an individual before they consider references. If you do supply names, check beforehand with those named, as a courtesy.

STANDARD RESUMÉ

SANDRA A. STUDENT

Present address (until April 30th, 2007):
500, University Avenue
London
ON, N6A 2V3
(519) 555-1234
e-mail: astudent@uwo.ca

Permanent address:
RR#5
Ilderton
ON, N0M 2A0
(519) 777-4321

Career Objectives: an entry-level position in Human Resources, where my background in organizational psychology would be an asset.

Education

B.A.(Hons) Psychology, University of Western Ontario (expected June 2007) (Honours Thesis: "Dilbert in Academia: Job Satisfaction among University Professors")

Honours and Awards

- Nobel Peace Prize, 2006
- Dean's Honours' List, 2004–2006
- President's Entrance Scholarship, 2004 ($5,000)

Work Experience

Summer 2005 – Research Assistant for Professor G. Humphrey Hall (Project title: "Social interactions among graduate students in different disciplines")
- helped develop questionnaire and conduct interviews
- assisted in statistical analysis of data
- drafted "Method" and "Results" sections of final report

Summer 2004 – Summer intern, United Nations, New York
- general filing and office help
- computer database maintenance
- assisted in negotiations to end all wars in the world

Summer 2003 – Sales associate and cashier, Wal-Mart, London

Specialized Skills and Experience

- Extensive computer skills. Knowledge of Windows XP; basic knowledge of Linux. Familiarity with the following programming languages: HTML, JavaScript, C++. Extensive experience with word-processing, database, spreadsheet, and presentation software.
- Custom designed Web pages for businesses in both Canada and the United States. Used knowledge of HTML, JavaScript, File Transfer Protocol, database, image and sound manipulation software.
- Strong statistical background. Took several advanced Statistics courses and was able to use these skills in my Research Assistant position.

Other Interests and Achievements

- Participated in competitive gymnastics at provincial level for 6 years. Member of Ontario Provincial Gymnastics Team for 3 years. Gymnastics coach. Assisted gymnasts aged 4–14 in advancing through the CanGym Levels System, while helping them gain new skills and confidence.
- Active interest in Theatre Arts. Member of Theatre Ontario, London Community Players, Theatre Western. Principal roles in several productions

References

Available upon request

A resumé with these traditional categories is not the only kind that works, however. Even if your experience is not directly related to the position you want, you can still write an effective *functional resumé*. Most functional resumés have categories for experience in different functions (for example, Research, Administration, Sales). Others may focus on personal attributes such as initiative, teamwork, analytic ability, or communication skills. If you choose to write a functional resumé, it's a good idea to include at the bottom a brief record of employment with dates, so that the reader has a firm grasp of when you did what.

FUNCTIONAL RESUMÉ

SANDRA B. STUDENT

RR#5
Granton
ON, NOM 1V0
(519) 777-4321
e-mail: xstudent@newserver.com

Career Objectives: Work in a public-relations or media organization dealing directly with clients or the public.

Profile: Excellent communicator with extensive experience working in front of an audience. Proven record of initiative.

Communications and public appearance experience

- Reporter for *The Gazette* newspaper at University of Western Ontario, Sept 2005–April 2006.
- Announcer on CHRW ("Radio Western") at University of Western Ontario, 2004–2005. Hosted weekly show "Awful Oldies," Jan–Dec 2005.
- Very active in Theatre Arts. Member of Theatre Ontario, London Community Players, Theatre Western. Principal roles in several productions.
- Perform regularly at local clubs as a singer/guitarist.

Initiative

- Founded my own business as a Web page designer for businesses in both Canada and the United States.
- Developed a Web site and bulletin board service for tropical bird owners to exchange information.
- Was able to support myself throughout University with contract Web design and computer programming jobs.

Education

B.A.(Hons) Psychology and English, University of Western Ontario (June 2001). At University, I specialized in industrial and organizational psychology, and in drama.

Other Achievements and Activities

- Shared the 2006 Nobel Peace Prize for my work with the United Nations on the elimination of global conflict.
- Participated in competitive gymnastics at the provincial level for 6 years. Member of Ontario Provincial Gymnastics Team for 3 years.

References

Available upon request

The tone of the resumé should be upbeat: don't draw attention to any potential weaknesses you might have, such as a lack of experience in a particular area. Never list a category and then write "None"—you don't want to suggest that you lack something! Remember to re-order your list of special skills to suggest a fit with each particular job you apply for, so that the reader will see at a glance that you have the needed skills. Finally, never claim more for yourself than is true; putting a falsehood into a resumé can be grounds for firing if it is discovered later.

You are not required to state anything about your age, place of birth, race, religion, or sex (use of initials, rather than your given names, may hide the last of these). Keep in mind, though, that if you are completing an application form with set questions, it is a good idea to provide all the information requested (especially about age and sex); if you don't, your application may be ignored.

PREPARING A LETTER OF APPLICATION FOR A JOB

Do not use the same letter for all applications: craft each one to focus directly on the particular job and company in question and to catch the attention of each particular reader. In a sense, both the resumé and the letter of application are intended to open the door to the next stage in the job hunt: the interview. The key is to link your skills to the position, not just to state information. What matters is not what *you want* but what the *employer needs*.

One challenge in writing a letter of application is to write about yourself and your qualifications without seeming egotistical. Two tips can help:

1. Limit sentences beginning with "I." Instead, try burying "I" in the middle of some sentences, where it will be less noticeable; for example, "For two months last summer, I worked as a . . .".

2. As much as you can, avoid making unsupported, subjective claims. Instead of saying, "I am a highly skilled manager," say something like, "Last summer, I managed a $50,000 field study with a crew of seven assistants." Rather than claim, "I have excellent research skills," you might say, "Based on my previous work, Professor Miriam Badani selected me from among ten applicants to help with her summer research work."

Here is an example (not to be copied rigidly), of an application letter that tries to connect the applicant's background with the needs of the company:

1 April 2002
Steven Nazar, Personnel Director
MegaDrug Corporation
110 Xenon Street
Toronto
ON,
M6Z 9Q1

Dear Mr. Nazar:

This letter is in response to your advertisement in the *Globe and Mail* on March 15 for a position as a Research Associate at MegaDrug Corporation. This entry-level position matches my career interests and is strongly compatible with my skills and experience. It is clear from the reports I have read that your company is well known as a leader in research into the development of new therapeutic agents, and I am very interested in becoming a member of such a highly regarded research team.

I will be graduating in May from Queen's University with a combined honours degree in physiology and psychology. The courses I have taken in these two disciplines have provided me with a strong background in basic human physiology as well as extensive experience in the design, running, and analysis of behavioural experiments involving both humans and animals. I have also had the opportunity to work as a Research Assistant for the last two summers in both the Physiology and Psychology Departments.

The enclosed resumé and university transcript will provide you with a more comprehensive account of my background, achievements, and interests.

I look forward to meeting with you to discuss the position.

Sincerely,

Sandra B. Student

FINAL WORDS OF ADVICE

When you apply for a job, your application is likely to be one of many. This means that it must pass an initial screening process before it is considered seriously. For that reason, it is *absolutely essential* that you submit a package that looks professional. First impressions count. Poor spelling, bad grammar, and a sloppy layout are guarantees that your application will end up in a wastepaper basket. Before you send out your application, go through the following checklist to be sure that everything is as it should be:

- ❑ Have I spell-checked both my resumé and my covering letter? Have I double-checked the spelling of the name of the person I am writing to on both the cover letter and the envelope?
- ❑ Have I made sure that there are no grammatical errors in my cover letter?
- ❑ Are both my resumé and my cover letter printed on good-quality paper?
- ❑ Does my resumé include everything about me that might be relevant to *this* job?
- ❑ Is the layout of my resumé neat, clear, and logically organized?

CHApter 16

COMMON ERRORS IN GRAMMAR AND USAGE

This chapter is not a comprehensive grammar lesson; it's simply a survey of those areas where students most often make mistakes. It will help you keep a lookout for weaknesses as you are editing your work. Once you get into the habit of checking your work, it won't be long before you are correcting potential problems as you write.

The grammatical terms used here are the most basic and familiar ones; if you need to review some of them, see Chapter 17 or Glossary II. For a thorough treatment of grammar and usage see Ruvinsky (2006).

TROUBLES WITH SENTENCE UNITY

SENTENCE FRAGMENTS

To be complete, a sentence must have both a subject and a verb in an independent clause; if it doesn't, it's a fragment. There are times in informal writing when it is acceptable to use a sentence fragment in order to emphasize a point, as in:

> ✓ What is the probability of contracting AIDS through casual contact? <u>Very low</u>.

In this example, the sentence fragment "Very low" is clearly intended to be understood as a short form of "The probability is very low." Unintentional sentence fragments, on the other hand, usually seem incomplete rather than shortened:

> ✗ The liquid was poured into a glass beaker. <u>Being a strong acid</u>.

The last "sentence" is incomplete because it has no subject or verb. (Remember that a participle such as *being* is a verbal, not a verb.) The sentence can be made complete by adding a subject and a verb:

> ✓ The <u>liquid</u> <u>was</u> a strong acid.

Alternatively, you could join the fragment to the preceding sentence:

✓ The liquid was poured into a glass beaker because it was a strong acid.

✓ Because the liquid was a strong acid, it was poured into a glass beaker.

RUN-ON SENTENCES

A run-on sentence is one that continues beyond the point where it should have stopped:

✗ The subjects who took part in the experiment said they enjoyed participating, even though it lasted for two hours, and they all agreed to come back for a second session, which will probably take place after reading week.

This run-on sentence could be fixed by removing the word "and" and by adding a period or semicolon after the word "hours."

Another kind of run-on sentence is one in which two independent clauses (phrases that could stand by themselves as sentences) are joined incorrectly by a comma:

✗ The instructions called for 50 g of sugar to be added, we used fructose in our experiment.

This error is known as a *comma splice*. There are three ways of correcting it:

1. by putting a period after "added" and starting a new sentence:

   ✓ . . . to be added. We . . .

2. by replacing the comma with a semicolon:

   ✓ . . . to be added; we . . .

3. by making one of the independent clauses subordinate to the other, so that it doesn't stand by itself:

   ✓ The instructions, which called for 50 g of sugar to be added, allowed us to use fructose in our experiment.

The one exception to the rule that independent clauses cannot be joined by a comma arises when the clauses are short and arranged in a tight sequence:

✓ I examined the data, I saw my mistake, and I changed my conclusion.

Contrary to what many people think, words such as *however, therefore,* and *thus* cannot be used to join independent clauses:

✗ Two of my friends started out in Chemistry, however they quickly decided they didn't like lab work.

The mistake can be corrected by beginning a new sentence after "Chemistry" or (preferably) by putting a semicolon in the same place:

✓ Two of my friends started out in Chemistry; however, they quickly decided they didn't like lab work.

The only words that can be used to join independent clauses are the coordinating conjunctions—*and, or, nor, but, for, yet,* and *so*—and subordinating conjunctions such as *if, because, since, while, when, where, after, before,* and *until*:

✓ Two of my friends started out in Chemistry, but they quickly decided they didn't like lab work.

FAULTY PREDICATION

When the subject of a sentence is not connected grammatically to what follows (the predicate), the result is *faulty predication*:

✗ The <u>reason</u> he failed was <u>because</u> he couldn't handle multiple-choice exams.

The problem here is that *the reason* means essentially the same thing as *because.* The subject needs a noun clause to complete it:

✓ The <u>reason</u> he failed was <u>that</u> he couldn't handle multiple-choice exams.

Another solution would be to rephrase the sentence:

✓ He failed because he couldn't handle multiple-choice exams.

Faulty predication also occurs with *is when* or *is where* constructions. These can be corrected in the same way:

✗ The difficulty <u>is when</u> the two sets of data disagree.

✓ The difficulty <u>arises when</u> the two sets of data disagree.

TROUBLES WITH SUBJECT-VERB AGREEMENT

IDENTIFYING THE SUBJECT

A verb should always agree in number with its subject. Sometimes, however, when the subject does not come at the beginning of the sentence or when it is separated from the verb by other information, you may be tempted to use a verb form that does not agree:

> ✗ The changes in the rate of flow <u>was measured</u> by the investigators.

The subject here is "changes," not "rate of flow"; therefore, the verb should be plural:

> ✓ The <u>changes</u> in the rate of flow <u>were measured</u> by the investigators.

EITHER, NEITHER, EACH

The indefinite pronouns *either*, *neither*, and *each* always take singular verbs:

> ✗ <u>Neither</u> of the cats <u>have</u> a flea collar.

> ✓ <u>Each</u> of them <u>has</u> a rabies tag.

COMPOUND SUBJECTS

When *or*, *either . . . or*, or *neither . . . nor* is used to create a compound subject, the verb should usually agree with the last item in the subject:

> ✓ Neither the professor nor <u>her students</u> <u>were able</u> to solve the equation.

> ✓ Either the students or <u>the TA</u> <u>was</u> misinformed.

You may find, however, that it sounds awkward in some cases to use a singular verb when a singular item follows a plural item:

> (✓) Either my chemistry <u>books</u> or my biology <u>text</u> <u>is</u> going to gather dust this weekend.

In such instances, it's better to rephrase the sentence:

> ✓ This weekend, I'm going to ignore either my chemistry books or my biology text.

The word *and* creates a compound subject and therefore takes a plural verb. *As well as* and *in addition to* do not create compound subjects, so the verb remains singular:

✓ Organic Chemistry <u>and</u> applied math <u>are</u> difficult subjects.

✓ Organic Chemistry <u>as well as</u> applied math <u>is</u> a difficult subject.

COLLECTIVE NOUNS

A collective noun is a singular noun that includes a number of members, such as *family, army*, or *team*. If the noun refers to the members as a unit, it takes a singular verb:

✓ The <u>class goes</u> on a field trip in June.

If, in the context of the sentence, the noun refers to the members as individuals, the verb becomes plural:

✓ The <u>team are receiving</u> their medals this week.

✓ The <u>majority</u> of bears <u>hibernate</u> in winter.

TITLES

The title of a book or a movie is always treated as a singular noun, even if it contains plural words; therefore it takes a singular verb:

✓ <u>*The Thirty Nine Steps* is</u> a fascinating book.

✓ <u>Bausch and Lomb is</u> a company that makes microscopes.

TENSE TROUBLES

Native speakers of English usually know without thinking which verb tense to use in a given context. However, a few tenses can still be confusing.

THE PAST PERFECT

If the main verb is in the past tense and you want to refer to something before that time, use the *past perfect* (*had* followed by the past participle). The time sequence will not be clear if you use the simple past in both clauses:

✗ He <u>hoped</u> that she <u>fixed</u> the printer.

✓ He <u>hoped</u> that she <u>had fixed</u> the printer.

Similarly, when you are reporting what someone said in the past—that is, when you are using past indirect discourse—you should use the past perfect tense in the clause describing what was said:

✗ He <u>told</u> the TA that he <u>wrote</u> the essay that week.

✓ He <u>told</u> the TA that he <u>had written</u> the essay that week.

USING "IF"

When you are describing a possibility in the future, use the present tense in the condition (*if*) clause and the future tense in the consequence clause:

✓ If she tests us on operant conditioning, I <u>will fail</u>.

When the possibility is unlikely, it is conventional—especially in formal writing—to use the subjunctive in the *if* clause, and *would* followed by the base verb in the consequence clause:

✗ If she <u>were to cancel</u> the test, I <u>would cheer</u>.

When you are describing a hypothetical instance in the past, use the past subjunctive (it has the same form as the past perfect) in the *if* clause and *would have* followed by the past participle for the consequence. A common error is to use *would have* in both clauses:

✗ If he <u>would have been</u> friendlier, I <u>would have asked</u> him to be my lab partner.

✓ If he <u>had been</u> friendlier, I <u>would have asked</u> him to be my lab partner.

PRONOUN TROUBLES

PRONOUN REFERENCE

The link between a pronoun and the noun it refers to must be clear. If the noun doesn't appear in the same sentence as the pronoun, it should appear in the preceding sentence:

✗ The <u>textbook supply</u> in the bookstore had run out, so we borrowed <u>them</u> from the library.

Since "textbook" is used as an adjective rather than a noun, it cannot serve as referent or antecedent for the pronoun "them." You must either replace "them" or change the phrase "textbook supply":

✓ The <u>textbook supply</u> in the bookstore had run out, so we borrowed <u>the texts</u> from the library.

✓ The bookstore had run out of <u>textbooks</u>, so we borrowed <u>them</u> from the library.

When a sentence contains more than one noun, make sure there is no ambiguity about which noun the pronoun refers to:

✗ The faculty want <u>increased salaries</u> as well as <u>fewer teaching hours</u>, but the administration does not favour <u>them</u>.

What does the pronoun *them* refer to: the salary increases, the reduced teaching hours, or both?

✓ The faculty want <u>increased salaries</u> as well as fewer teaching hours, but the administration does not favour <u>salary increases</u>.

USING "IT" AND "THIS"
Using *it* and *this* without a clear referent can lead to confusion:

✗ Although the directors wanted to meet in January, <u>it</u> (<u>this</u>) didn't take place until May.

✓ Although the directors wanted to meet in January, <u>the conference</u> didn't take place until May.

✓ Although the directors wanted to schedule <u>the meeting</u> for January, <u>it</u> (<u>this</u>) didn't take place until May.

Make sure that *it* or *this* clearly refers to a specific noun or pronoun.

PRONOUN AGREEMENT AND GENDER
A pronoun should agree in number and person with the noun that it refers to. However, an increasing awareness of sexist or biased language has changed what is considered acceptable over the last few decades. In the past, the following sentence would have been considered incorrect:

When <u>a student</u> is sick, <u>their</u> classmates usually help out.

It would have been corrected to read:

When <u>a student</u> is sick, <u>his</u> classmates usually help out.

This is because, traditionally, the word *his* has been used to indicate both male and female. Although many grammarians still maintain that *he* and *his* have dual meanings—one for an individual male and one for any human being—today this usage is increasingly regarded as sexist. As we pointed out in Chapter 14, an alternative to using *his* is to use *his or her*, as this handbook occasionally does, or *he/she*; however, these phrases are awkward, particularly if they have to be used repeatedly in a passage. For this reason, using *their* or *they* to indicate an individual of either gender is becoming more and more common, and this trend appears to be gaining acceptance. As a result, the first sentence above would now be considered correct by many, though there are those who would still find it objectionable. Where possible, it's better to rephrase the sentence—in this case, by switching from the singular to the plural in both the noun and the pronoun:

> ✓ When <u>students</u> are sick, <u>their</u> classmates usually help out.

USING "ONE"

People often use the word *one* to avoid overusing *I* in their writing. Although in Britain this is common, in Canada and the United States frequent use of *one* may seem too formal and even a bit pompous:

> If <u>one</u> were to apply for the grant, <u>one</u> would find <u>oneself</u> engulfed in so many bureaucratic forms that <u>one's</u> patience would be stretched thin.

In the past, a common way around this problem was to use the third-person *his* or (less often) *her* as the adjectival form of *one*:

> <u>One</u> would find <u>his</u> (or <u>her</u>) patience stretched thin.

Today, this usage is regarded with less favour. As we saw in the preceding section, you may be able to substitute the plural *their*; just remember that some people still object to this usage as well. The best solution, again, may be to rephrase the sentence with a plural subject:

> ✓ If <u>researchers</u> were to apply for grants, <u>they</u> would find <u>themselves</u> engulfed in so many bureaucratic forms that <u>their</u> patience would be stretched thin.

In any case, try to use *one* sparingly, and don't be afraid of the occasional *I*. The one serious error to avoid is mixing the third-person *one* with the second-person *you*.

> ✗ When <u>one</u> visits the cyclotron, <u>you</u> are impressed by its size.

In formal academic writing generally, *you* is not an appropriate substitute for *one*.

USING "ME" AND OTHER OBJECTIVE PRONOUNS

Remembering that it's wrong to say "Sherra and me were invited to present our findings to the delegates" rather than "Sherra and I were invited . . .", many people use the subjective form of the pronoun even when it should be objective:

✗ The delegates invited Sherra and I to present our findings.

✓ The delegates invited Sherra and me to present our findings.

The verb "invited" requires an object; "me" is in the objective case. A good way to tell which form is correct is to ask yourself how the sentence would sound if the pronoun were used by itself. It will be obvious that the subjective form—"The delegates invited I . . ."—is not appropriate.

The same problem often arises with prepositions, which should also be followed by a noun in the objective case:

✗ Between you and I, this result doesn't make sense.

✓ Between you and me, this result doesn't make sense.

✗ Eating well is a problem for we students.

✓ Eating well is a problem for us students.

There are times, however, when the correct case can sound stiff or awkward:

(✓) To whom was the award given?

Rather than keeping to a correct but awkward form, try to reword the sentence:

✓ Who received the award?

EXCEPTIONS FOR PRONOUNS FOLLOWING PREPOSITIONS

The rule that a pronoun following a preposition takes the objective case has exceptions. When the preposition is followed by a clause, the pronoun should take the case required by its position in the clause:

✗ The students showed some concern about whom would be selected as Dean.

Although the pronoun follows the preposition "about," it is also the subject of the verb "would be selected" and therefore requires the subjective case:

✓ The students showed some concern <u>about</u> <u>who would be selected as Dean</u>.

Similarly, when a gerund (a word that acts partly as a noun and partly as a verb) is the subject of a clause, the pronoun that modifies it takes the possessive case:

✗ We were surprised <u>by</u> <u>him dropping</u> out of school.

✓ We were surprised <u>by</u> <u>his dropping</u> out of school.

TROUBLES WITH MODIFYING

Adjectives modify nouns; adverbs modify verbs, adjectives, and other adverbs. Do not use an adjective to modify a verb:

✗ He played <u>good</u>. (adjective with verb)

✓ He played <u>well</u>. (adverb modifying verb)

✓ He played <u>really</u> <u>well</u>. (adverb modifying adverb)

✓ He had a <u>good</u> <u>style</u>. (adjective modifying noun)

✓ He had a <u>really</u> <u>good</u> style. (adverb modifying adjective)

SQUINTING MODIFIERS

Remember that clarity depends largely on word order: to avoid confusion, the relations between the different parts of a sentence must be clear. Modifiers should therefore be as close as possible to the words they modify. A *squinting modifier* is one that, because of its position, seems to look in two directions at once:

✗ She discovered <u>in the spring</u> she was going to have to write final exams.

Is "spring" the time of the discovery or the time of the exams? The logical relation is usually clearest when you place the modifier immediately before or after the element it modifies:

✓ <u>In the spring</u> she <u>discovered</u> that she was going to have to write final exams.

✓ She discovered that she would have to write <u>final exams</u> <u>in the spring</u>.

Other squinting modifiers can be corrected in the same way:

✗ Our biology professor gave a lecture on *Planaria*, <u>which was well illustrated</u>.

✓ Our biology professor gave a <u>well-illustrated</u> <u>lecture</u> on *Planaria*.

DANGLING MODIFIERS

Modifiers that have no grammatical connection with anything else in the sentence are said to be dangling:

✗ <u>Walking</u> around the campus in June, the river and trees made a picturesque scene.

Who is doing the walking? Here's another example:

✗ <u>Reflecting</u> on the results of the study, it was decided not to submit the paper for publication.

Who is doing the reflecting? Clarify the meaning by connecting the dangling modifier to a new subject:

✓ <u>Walking</u> around the campus in June, <u>Imelda</u> thought the river and trees made a picturesque scene.

✓ <u>Reflecting</u> on the results of the study, <u>the research team</u> decided not to submit the paper for publication.

TROUBLES WITH PAIRS AND PARALLELS

COMPARISONS

Make sure that your comparisons are complete. The second element in a comparison should be equivalent to the first, whether the equivalence is stated or merely implied.

✗ Today's students have a greater understanding of calculus than their parents.

The sentence suggests that the two things being compared are "calculus" and "parents." Adding a second verb ("have") that is equivalent to the first one shows that the two things being compared are parents' understanding and students' understanding:

> ✓ Today's students <u>have</u> a greater understanding of calculus than their parents <u>have</u>.

A similar problem arises in the following comparison:

> ✗ The new text is a <u>boring book</u> and so are the lectures.

The lectures may be boring, but they are not a "boring book"; to make sense, the two parts of the comparison must be parallel:

> ✓ The new text is <u>boring</u> and so are the lectures.

CORRELATIVES (COORDINATE CONSTRUCTIONS)

Constructions such as *both . . . and, not only . . . but,* and *neither . . . nor* are especially tricky. The coordinating term must not come too early or else one of the parts that come after will not connect with the common element. For the implied comparison to work, the two parts that come after the coordinating term must be grammatically equivalent:

> ✗ He <u>not only</u> studies music <u>but also</u> math.

> ✓ He studies <u>not only</u> music <u>but also</u> math.

PARALLEL PHRASING

A series of items in a sentence should be phrased in parallel wording. Make sure that all the parts of a parallel construction are really equal:

> ✗ We had to turn in <u>our</u> rough notes, <u>our</u> calculations, and finished assignment.

> ✓ We had to turn in <u>our</u> rough notes, <u>our</u> calculations, and <u>our</u> finished assignment.

Once you have decided to include the pronoun "our" in the first two elements, the third must have it too.

For clarity as well as stylistic grace, keep similar ideas in similar form:

> ✗ He <u>failed</u> Genetics and <u>barely passed</u> Statistics, but Zoology <u>was</u> a subject he did well in.

> ✓ He <u>failed</u> Genetics and <u>barely passed</u> Statistics, but <u>did well</u> in Zoology.

REFERENCE

Ruvinsky, M. (2006). *Practical grammar: A Canadian writer's resource*. Don Mills, ON: Oxford University Press.

chApter 17

PUNCTUATION

Punctuation causes students so many problems that it deserves a chapter of its own. If your punctuation is faulty, your readers will be confused and may have to backtrack; worse still, they may be tempted to skip over the rough spots. Punctuation marks are the traffic signals of writing: use them with precision to keep readers moving smoothly through your work. Most of the rules we give below are general, but when there were several possibilities (for example, when British and American usage differs), we have used the APA style guidelines.

Items in this chapter are arranged alphabetically: *apostrophe, brackets, colon, comma, dash, ellipsis, exclamation mark, hyphen, italics, parentheses, period, quotation marks,* and *semicolon*.

APOSTROPHE [']

1. **Use an apostrophe to indicate possession. The following rules are the easiest to remember:**

   a) To illustrate the possessive, create an "of" phrase:

the Perkins house	⟶	the house *of the Perkins*
the girls fathers	⟶	the fathers *of the girls*
the childrens parents	⟶	the parents *of the children*
Shakespeares plays	⟶	the plays *of Shakespeare*

   b) If the noun in the "of" phrase ends in "s", add an apostrophe:

   the Perkins' house
   the girls' fathers

   c) If the noun in the "of" phrase does not end in "s", add an apostrophe plus "s":

   the children's parents
   Shakespeare's plays

2. **Use an apostrophe to show contractions of words:**

   isn't   we'll   he's   shouldn't   I'm

Caution: don't confuse *it's*, the contraction of "it is," with *its*, the possessive of "it," which has no apostrophe. And remember that prossessive pronouns *never* take an apostrophe: yours, hers, its, ours, theirs

BRACKETS [ ]

Brackets are square enclosures, not to be confused with parentheses (which are round).

Use brackets to set off a remark of your own within a quotation. They show that the words enclosed are not those of the person quoted:

> As Mitchell (1984) stated, "Several of the changes observed [in the cat] are seen also in the monkey."

Brackets are sometimes used to enclose *sic*, which is used after an error, such as a misspelling, to show that the mistake was in the original. *Sic* may be italicized or underlined:

> In describing the inhabitants of a tidal pool, he wrote that "it was almost impossible to loosen barnikles [*sic*] from the rock surface."

COLON [:]

A colon indicates that something is to follow.

1. **Use a colon before a formal statement or series:**

   ✓ The layers are the following: sclera, choroid, and retina.

   Do not use a colon if the words preceding it do not form a complete sentence:

   ✗ The layers are: sclera, choroid, and retina.

   ✓ The layers are sclera, choroid, and retina.

2. **Use a colon for formality before a direct quotation:**

   The instructor was adamant: "All students must take the exam today."

COMMA [,]

Commas are the trickiest of all punctuation marks: even the experts differ on when to use them. Most people agree, however, that too many commas are as bad as too few, since they make writing choppy and awkward to read. Certainly recent writers use fewer commas than earlier stylists did. Whenever you are in doubt, let clarity be your guide. The most widely accepted conventions are these:

1. **Use a comma to separate two independent clauses joined by a coordinating conjunction (and, but, for, or, nor, yet, so).** By signalling that there are two clauses, the comma will prevent the reader from confusing the beginning of the second clause with the end of the first:

    ✗ He finished working with the microscope and his partner turned off the power.

    ✓ He finished working with the microscope, and his partner turned off the power.

    When the second clause has the same subject as the first, you have the option of omitting both the second subject and the comma:

    ✓ She writes well, but she never finishes on time.

    ✓ She writes well but never finishes on time.

    If you mistakenly punctuate two sentences as if they were one, the result will be a *run-on sentence*; if you use a comma but forget the coordinating conjunction, the result will be a *comma splice*:

    ✗ He took his class to the zoo, it was closed for repairs.

    ✓ He took his class to the zoo, but it was closed for repairs.

    Remember that words such as *however, therefore,* and *thus* are conjunctive adverbs, not conjunctions: if you use one of them the way you would use a conjunction, the result will again be a *comma splice*:

    ✗ She was accepted into medical school, however, she took a year off to earn her tuition.

    ✓ She was accepted into medical school; however, she took a year off to earn her tuition.

Conjunctive adverbs are often confused with conjunctions. You can distinguish between the two if you remember that a conjunctive adverb's
position in a sentence can be changed:

✓ She was accepted into medical school; she took a year off, <u>however</u>, to
earn her tuition.

The position of a conjunction, on the other hand, is invariable; it must be
placed between the two clauses:

✓ She was accepted into medical school, <u>but</u> she took a year off to earn
her tuition.

When, in rare cases, the independent clauses are short and closely related,
they may be joined by a comma alone:

✓ I came, I saw, I conquered.

A *fused sentence* is a run-on sentence in which independent clauses are
slapped together with no punctuation at all:

✗ He watched the hockey game all afternoon the only exercise he got
was going to the kitchen between periods.

A fused sentence sounds like breathless babbling—and it's a serious
error.

2. **Use a comma between items in a series.** Place a coordinating conjunction before the last item:

✓ The room that housed the animals was large, bright, and clean.

✓ There is a cage-washer, a bottle-washer, and a place for storing the
clean glassware.

The comma before the conjunction is optional:

✓ We have an office, a lab and a surgery.

Sometimes, however, the final comma can help to prevent confusion:

✗ We arranged to move the rats, photographs of the lab and the gerbil
food.

In this case, a comma can prevent the reader from thinking that "photographs" refers both to "the lab" and to "the gerbil food":

✓ We arranged to move the rats, photographs of the lab, and the gerbil food.

3. **Use a comma to separate adjectives preceding a noun when they modify the same element:**

   ✓ It was a reliable, accurate weighing device.

When the adjectives do not modify the same element, you should not use a comma:

✗ It was an expensive, chemical balance.

Here "chemical" modifies "balance," but "expensive" modifies the whole phrase "chemical balance." A good way of checking whether or not you need a comma is to see if you can reverse the order of the adjectives. If you can reverse it (*reliable, accurate balance* or *accurate, reliable balance*), use a comma; if you can't (*chemical expensive balance*), omit the comma:

✓ It was an expensive chemical balance.

4. **Use commas to set off an interruption (or "parenthetical element"):**

   ✓ The outcome, he said, was a complete failure.

   ✓ My tutor, however, couldn't answer the question.

Remember to put commas on *both sides* of the interruption:

✗ My tutor however, couldn't answer the question.

✗ The equipment, they reported was obsolete.

✓ The equipment, they reported, was obsolete.

5. **Use commas to set off words or phrases that provide additional but non-essential information:**

   ✓ Her grade in statistics, her favourite course, was not very high.

   ✓ The new computer, his pride and joy, was always crashing.

In these examples, "her favourite course" and "his pride and joy" are appositives: they give additional information about the nouns they refer to ("statistics" and "computer"), but the sentences would be understandable without them. Here's another example:

✓ *Equinox* magazine, which is published locally, often contains material that I can use in my course.

The phrase "which is published locally" is called a *non-restrictive modifier* because it does not limit the meaning of the words it modifies ("*Equinox* magazine"). Without that modifying clause the sentence would still refer to the contents of the magazine. Since the information the clause provides is not necessary to the meaning of the sentence, you must use commas on both sides to set it off.

In contrast, a *restrictive modifier* is one that provides essential information; therefore it must not be set apart from the element it modifies, and commas should not be used:

✓ The magazine that has the black cover is *Equinox*.

Without the clause "that has the black cover," the reader would not know which magazine was *Equinox*. To avoid confusion, be sure to distinguish carefully between essential and additional information. The difference can be important:

Students who are not willing to work should not receive grants.

Students, who are not willing to work, should not receive grants.

6. **Use a comma after an introductory phrase when omitting it would cause confusion:**

✗ In the room behind the students flew paper airplanes.

✓ In the room behind, the students flew paper airplanes.

7. **Use a comma to separate elements in dates and addresses:**

February 2, 2007. (Commas are often omitted if the day comes first: 2 February 2007.)

117 Hudson Drive, Edmonton, Alberta.

They lived in Dartmouth, Nova Scotia.

8. **Use a comma before a quotation in a sentence:**

    ✓ He stated, "*E. coli* was the bacterium isolated."

    ✓ "The most difficult part of the procedure," she reported, "was finding the material to work with."

    For more formality, you may use a colon (see page 189).

9. **Use a comma with a name followed by a title:**

    D. Gunn, Ph.D.

    Patrice Lareau, M.D.

DASH [—]

A dash creates an abrupt pause, emphasizing the words that follow. Never use dashes as casual substitutes for other punctuation: overuse can detract from the calm, well-reasoned effect you want to create.

1. **Use a dash to stress a word or phrase:**

    The fire alarm—which was deafening—warned them of the danger.

    I thought that writing this paper would be easy—when I started.

2. **Use a dash in interrupted or unfinished dialogue:**

    "It's a matter—to put it delicately—of personal hygiene."

    "But I thought—" Donald began, but Mario cut him off: "You were wrong."

    In typing, use two hyphens together, with no spaces on either side, to show a dash. Your word processor may automatically convert this to a dash for you as you continue typing.

ELLIPSIS [. . .]

1. **Use an ellipsis (three spaced dots) to show an omission from a quotation:**

    "The hormonal control of reproduction is modulated . . . ultimately by the production of gonadal steroids."

If the omission comes at the beginning of the quotation, an ellipsis is not necessarily used:

> She explained that reproductive control is modulated "by the production of gonadal steroids."

If the omission comes at the end of a sentence, the ellipsis is followed by a fourth period:

> She explained that "control of reproduction is modulated. . . ."

2. **Use an ellipsis to show that a series of numbers continues indefinitely:**

> 1, 3, 5, 7, 9 . . .

EXCLAMATION MARK [!]

An exclamation mark helps to show emotion or feeling. It is usually found in dialogue:

> "Woe is me!" she cried.

In scientific writing, there is virtually no time when you would need to use it.

HYPHEN [-]

1. **Use a hyphen if you must divide a word at the end of a line.** When a word is too long to fit at the end of a line, it's better to start a new line than to break the word. If you must divide, however, remember these rules:

- Divide between syllables.
- Never divide a one-syllable word.
- Never leave one letter by itself.
- Divide double consonants except when they come before a suffix, in which case divide before the suffix:

> ar-rangement
> embar-rassment
> fall-ing
> pass-able

When the second consonant has been added to form the suffix, keep it with the suffix:

refer-ral
begin-ning

2. **Use a hyphen to separate the parts of certain compound words:**

   - compound nouns:

     test-tube; vice-president

   - compound verbs:

     fine-tune; test-drive

   - compound nouns and adjectives used as modifiers preceding nouns:

     a well-designed study; sixteenth-century science

   When you are not using such expressions adjectivally, do not hyphenate them:

     The study was well designed.

     It dates from the sixteenth century.

   Most hyphenated nouns and verbs lose the hyphen over time. When in doubt, check a dictionary.

3. **Use a hyphen with certain prefixes (*all-*, *self-*, *ex-*, and those prefixes preceding a proper name):**

     all-trans retinal; self-imposed; ex-student; pro-nuclear; trans-Canada

4. **Use a hyphen to emphasize contrasting prefixes:**

     Both pre- and post-treatment measures were taken.

5. **Use a hyphen to separate written-out compound numbers from one to a hundred and compound fractions.** Remember that numbers above nine are written only when they begin a sentence:

     Eighty-one centimetres; seven-tenths full

6. **Use a hyphen to separate parts of inclusive numbers or dates:**

> the years 1973-1976; pages 3-40

ITALICS [*ITALICS*]

Italics are slanted (cursive) letters. There are several occasions when you would want to use italics:

1. **titles of books and periodicals:**

> Darwin's *Origin of Species*

2. **biological names:**

> the common or brown rat, *Rattus norvegicus*

3. **when a word or phrase is used as a linguistic example:**

> They were asked to solve an anagram of the word *caveat*.

4. **words that could be misread or misunderstood:**

> the *aging* professors (those studying older people)

5. **letters used as symbols or algebraic terms:**

> *SD* (standard deviation)
>
> $df = 17$ (degrees of freedom)

You can also use italics for emphasizing a word or an idea. However, you should not do this too often, or the emphasis will lose its force.

PARENTHESES [( )]

1. **Use parentheses to enclose an explanation, example, or qualification.** Parentheses show that the enclosed material is of incidental importance to the main idea. They make an interruption that is more subtle than one marked off by dashes but more pronounced than one set off by commas:

> The meerkat (a mongoose-like animal) is found in southern Africa.
>
> At least 30 people (according to the newspaper report) were under observation.

Remember that punctuation should not precede parentheses, but it may follow them if required by the sense of the sentence:

> There were some complaints (mainly from the less experienced students), but we decided to continue with the project anyway.

If the parenthetical statement comes between two complete sentences, it should be punctuated as a sentence, with the period, question mark, or exclamation mark inside the parentheses:

> I finished my last essay on April 3. (It was on long-term memory.) Fortunately, I had three weeks free to study for the exam.

2. **Use parentheses to enclose reference citations.** See Chapter 11 for details.

PERIOD [.]

1. **Use a period at the end of a sentence.** A period indicates a full stop, not just a pause.

2. **Use a period with abbreviations.** British style omits the period in certain cases, but North American style usually requires it for abbreviated titles (Mrs., Dr., Rev., Ph.D., etc.), as well as for place names (B.C., N.W.T., P.E.I., N.Y., D.C.); however, in official postal use, two-letter state and provincial abbreviations do not require periods (BC, NT, PE, NY, DC). Although the abbreviations and acronyms for some organizations include periods, the most common ones generally do not (APA, CARE, CIDA, CBC, RCMP, etc.).

3. **Use a period at the end of an indirect question.** Do *not* use a question mark:

   ✗ He asked if I wanted a clean lab coat?

   ✓ He asked if I wanted a clean lab coat.

QUOTATION MARKS [" " OR ' ']

Quotation marks are used for several purposes, as described below. Conventions regarding single or double quotation marks vary, but APA style recommends that you use double marks for quotations within the running text and single for quotations within quotations.

1. **Use quotation marks to signify direct discourse (the actual words of a speaker):**

   > I asked, "What is the matter?"

   > "I have a pain in my big toe," he replied.

2. **Use quotation marks to show that words themselves are the issue:**

   > The term "information processing" has a distinct meaning in psychology.

   Alternatively, you may italicize or underline the terms in question.
   Sometimes quotation marks are used to mark a slang word or an inappropriate usage to show that the writer is aware of the difficulty:

   > Several of the "experts" did not seem to know anything about the topic.

   Use this device only when necessary. In general, it's better to let the context show your attitude, or to choose another term.

3. **Use single quotation marks to enclose quotations within quotations:**

   > He said, "several of the 'experts' did not seem to know anything about the topic."

PLACEMENT OF PUNCTUATION WITH QUOTATION MARKS

The *APA manual* suggests the following guidelines:

- Periods and commas always go inside closing quotation marks:

   > He said, "I think we can finish tonight," but I told him, "Hector, it's time to go home."

- A semicolon or colon always goes outside the quotation marks:

   > Conrad calls it "a masterpiece"; I call it junk.

- A question mark, dash, or exclamation mark goes inside the quotation marks if it is part of the quotation, but outside if not:

    She asked, "What *is* that, Conrad?"

    Did she really call it "a piece of junk"?

    You could hardly call it "a masterpiece"!

    I was just telling Louisa, "*I* think it looks like—" when Conrad walked in the room.

- With a block quotation (one that is set off from the running text), you do not need to use any quotation marks, but for quotations within the block use double quotation marks.

SEMICOLON [;]

1. **Use a semicolon to join independent clauses (complete sentences) that are closely related:**

    For five days he worked non-stop; by Saturday he was exhausted.

    His lecture was confusing; no one could understand the terminology.

    A semicolon is especially useful when the second independent clause begins with a conjunctive adverb such as *however, moreover, consequently, nevertheless, in addition*, or *therefore* (usually followed by a comma):

    He made several attempts; however, none of them was successful.

    It's usually acceptable to follow a semicolon with a coordinating conjunction if the second clause is complicated by other commas:

    Some of these animals, wolverine and lynx in particular, are rarely seen; but occasionally, if you are patient, you might catch a glimpse of one.

2. **Use a semicolon to mark the divisions in a complicated series when individual items themselves need commas.** Using a comma to mark the subdivisions and a semicolon to mark the main divisions will help to prevent mix-ups:

    ✗   He invited Prof. Ludvik, the vice-principal, Christine Li, and Dr. J. Dexter Schokmann.

Is the vice-principal a separate person?

✓ He invited Prof. Ludvik; the vice-principal, Christine Li; and Dr. J. Dexter Schokmann.

In a case such as this, the elements separated by the semicolon need not be independent clauses.

CHApter 18

CATCHLIST OF MISUSED WORDS AND PHRASES

This chapter offers a catchlist of words and phrases that are often misused. A periodic read-through will refresh your memory and help you avoid needless mistakes.

accept, except. Accept is a verb meaning to *receive affirmatively*; **except**, when used as a verb, means to *exclude*:

> I <u>accept</u> your offer.

> The teacher <u>excepted</u> him from the general punishment.

accompanied by, accompanied with. Use **accompanied by** for people; use **accompanied with** for objects:

> He was <u>accompanied by</u> his wife.

> The brochure arrived, <u>accompanied with</u> a discount coupon.

advice, advise. Advice is a noun, **advise** a verb:

> He was <u>advised</u> to ignore the others' <u>advice</u>.

affect, effect. Affect is a verb meaning to *influence*; however, it also has a specialized meaning in psychology, referring to a person's emotional state. **Effect** can be either a noun meaning *result* or a verb meaning to *bring about*:

> The eye drops <u>affect</u> his vision.

> Because he was so depressed, he showed no <u>affect</u> when he heard the joke.

> The <u>effect</u> of higher government spending is higher inflation.

> People lack confidence in their ability to <u>effect</u> change in society.

all ready, already. To be **all ready** is simply to be ready for something; **already** means *beforehand* or *earlier*:

> The students were all ready for the lecture to begin.
>
> The professor had already left her office by the time Blair arrived.

all right. Write as two separate words: *all right*. This can mean *safe and sound, in good condition, okay*; *correct*; *satisfactory*; or *I agree*:

> Are you all right?
>
> The student's answers were all right.

(Note the ambiguity of the second example: does it mean that the answers were all correct or simply satisfactory? In this case, it might be better to use a clearer word.)

all together, altogether. **All together** means *in a group*; **altogether** is an adverb meaning *entirely*:

> He was altogether certain that the children were all together.

allusion, illusion. An **allusion** is an indirect reference to something; an **illusion** is a false perception:

> The rock image is an allusion to the myth of Sisyphus.
>
> He thought he saw a sea monster, but it was an illusion.

a lot. Write as two separate words: *a lot*.

alternate, alternative. **Alternate** means *every other* or *every second* thing in a series; **alternative** refers to a *choice* between options:

> The two sections of the class attended discussion groups on alternate weeks.
>
> The students could do an extra paper as an alternative to writing the exam.

among, between. Use **among** for three or more persons or objects, **between** for two:

> Between you and me, there's trouble among the team members.

amount, number. **Amount** indicates quantity when units are not discrete and not absolute; **number** indicates quantity when units are discrete and absolute:

> A large <u>amount</u> of timber.

> A large <u>number</u> of students.

See also **less, fewer.**

analysis. The plural is **analyses.**

anyone, any one. **Anyone** is written as two words to give numerical emphasis; otherwise it is written as one word:

> <u>Any one</u> of us could do that.

> <u>Anyone</u> could do that.

anyways. Non-standard English: use *anyway*.

as, because. **As** is a weaker conjunction than **because** and may be confused with *when*:

> ✗ <u>As</u> I was working, I ate at my desk.

> ✓ <u>Because</u> I was working, I ate at my desk.

> ✗ He arrived <u>as</u> I was leaving.

> ✓ He arrived <u>when</u> I was leaving.

as to. A common feature of bureaucratese. Replace it with a single-word preposition such as *about* or *on*:

> ✗ They were concerned <u>as to</u> the range of disagreement.

> ✓ They were concerned <u>about</u> the range of disagreement.

> ✗ They recorded his comments <u>as to</u> the treaty.

> ✓ They recorded his comments <u>on</u> the treaty.

bad, badly. **Bad** is an adjective meaning *not good*:

> The meat <u>tastes bad</u>.

> He <u>felt bad</u> about forgetting the dinner party.

Badly is an adverb meaning *not well*; when used with the verbs **want** or **need,** it means *very much*:

> She thought he played the villain's part badly.

> I badly need a new suit.

beside, besides. Beside is a preposition meaning *next to*:

> She worked beside her assistant.

Besides has two uses: as a preposition it means *in addition to*; as a conjunctive adverb it means *moreover*:

> Besides recommending the changes, the consultants are implementing them.

> Besides, it was hot and we wanted to rest.

between. See **among**.

bring, take. One **brings** something to a closer place and **takes** it to a farther one:

> Take it with you when you go.

> Next time you come to visit, bring your friend along.

can, may. Can means to *be able*; **may** means to *have permission*:

> Can you fix the lock?

> May I have another piece of cake?

In speech, **can** is used to cover both meanings; in formal writing, however, you should observe the distinction.

can't hardly. A faulty combination of the phrases **can't** and **can hardly.** Use one or the other:

> He can't swim.

> He can hardly swim.

cite, sight, site. To **cite** something is to *quote* or *mention* it as an example or authority; **sight** can be used in many ways, all of which relate to the ability to *see*; **site** refers to a specific *location,* a particular place at which something is located.

complement, compliment. The verb to **complement** means to *complete* or *enhance;* to **compliment** means *to praise*:

> Her ability to analyze data complements her excellent research skills.

> I complimented her on her outstanding report.

The same rule applies when these words are used as adjectives:

> Blue and yellow are complimentary colours.

> They gave me complimentary tickets.

compose, comprise. Both words mean to *constitute* or *make up,* but **compose** is preferred. **Comprise** is correctly used to mean *include, consist of,* or *be composed of.* Using **comprise** in the passive ("is comprised of")—as you might be tempted to do in the second example below—is usually frowned on in formal writing:

> These students compose the group which will go overseas.

> Each paragraph comprises an introduction, an argument, and a conclusion.

continual, continuous. **Continual** means *repeated over a period of time;* **continuous** means *constant* or *without interruption*:

> The strikes caused continual delays in building the road.

> Five days of continuous rain ruined our holiday.

could of. This construction is incorrect, as are **might of, should of,** and **would of.** Replace *of* with *have:*

> ✗ He could of done it.
> ✓ He could have done it.
> ✓ They might have been there.
> ✓ I should have known.
> ✓ We would have left earlier.

council, counsel. Council is a noun meaning an *advisory* or *deliberative assembly*. **Counsel** as a noun means *advice* or *lawyer*; as a verb it means to *give advice*.

> The college council meets on Tuesday.

> We respect her counsel, since she's seldom wrong.

> As a camp counsellor, you may need to counsel parents as well as children.

criterion, criteria. A **criterion** is a standard for judging something. **Criteria** is the plural of **criterion** and thus requires a plural verb:

> These are my criteria for grading the reports.

data. The plural of **datum**. The set of information, usually in numerical form, that is used for analysis as the basis for a study. Informally, **data** is often used as a singular noun, but in formal contexts it should be treated as a plural:

> These data were gathered in an unsystematic fashion.

Because *data* often refers to a single mass entity, many writers now accept its use with a single verb and pronoun:

> When the data is in we'll review it.

deduce, deduct. To **deduce** something is to *work it out by reasoning*; to **deduct** means to *subtract* or *take away* from something. The noun form of both words is **deduction**.

defence, defense. Both spellings are correct: **defence** is standard in Britain and is somewhat more common in Canada; **defense** is standard in the United States.

delusion, illusion. A delusion is a belief or perception that is distorted; an **illusion** is a false belief:

> He had delusions of grandeur.

> The desert pool he thought he saw was an illusion.

dependent, dependant. **Dependent** is an adjective meaning *contingent on* or *subject to*; **dependant** is a noun.

> Suriya's graduation is <u>dependent</u> upon her passing algebra.

> Chedley is a <u>dependant</u> of his father.

device, devise. The word ending in **-ice** is the noun; the word ending in **-ise** is the verb.

different than. Use **different from** to compare two persons or things.

> You are <u>different from</u> me.

diminish, minimize. To **diminish** means to *make* or *become smaller*; to **minimize** means to *reduce* something to the smallest possible amount or size.

disinterested, uninterested. **Disinterested** implies impartiality or neutrality; **uninterested** implies a lack of interest:

> As a <u>disinterested</u> observer, he was in a good position to judge the issue fairly.

> <u>Uninterested</u> in the proceedings, he yawned repeatedly.

due to. Although increasingly used to mean *because of*, **due** is an adjective and therefore needs to modify something:

> ✗ <u>Due to</u> his impatience, we lost the contract. [<u>Due</u> is dangling.]

> ✓ The loss was <u>due to</u> his impatience.

e.g., i.e. **E.g.** means *for example*; **i.e.** means *that is*. It is incorrect to use them interchangeably.

entomology, etymology. **Entomology** is the study of insects; **etymology** is the study of the derivation and history of words.

exceptional, exceptionable. **Exceptional** means *unusual* or *outstanding*, whereas **exceptionable** means *open to objection* and is generally used in negative contexts.

> His accomplishments are <u>exceptional</u>.

> He was ejected from the game because of his <u>exceptionable</u> behaviour.

farther, further. **Farther** refers to distance, **further** to extent:

> He paddled farther than his friends.

> She explained the plan further.

focus. The plural of the noun may be either **focuses** (also spelled **focusses**) or **foci.**

good, well. **Good** is an adjective that modifies a noun; **well** is an adverb that modifies a verb.

> He is a good rugby player.

> The experiment went well.

hanged, hung. **Hanged** means *executed by hanging.* **Hung** means *suspended* or *clung to*:

> He was hanged at dawn for the murder.

> He hung the picture.

> She hung on to the boat when it capsized.

hereditary, heredity. **Heredity** is a noun; **hereditary** is an adjective. **Heredity** is the biological process whereby characteristics are passed from one generation to the next; **hereditary** describes those characteristics:

> Heredity is a factor in the incidence of this desease.

> Your asthma may be hereditary.

hopefully. Use **hopefully** as an adverb meaning *full of hope*:

> She scanned the horizon hopefully, looking for signs of the missing boat.

In formal writing, using **hopefully** to mean *I hope* is still frowned upon, although it is increasingly common; it's better to use *I hope*:

> ✗ Hopefully the experiment will go off without a hitch.

> ✓ I hope the experiment will go off without a hitch.

i.e. This is *not* the same as **e.g.** See **e.g.**

illusion. See **delusion.**

incite, insight. Incite is a verb meaning to *stir up*; **insight** is a noun meaning (often sudden) understanding.

infer, imply. To **infer** means to *deduce* or *conclude by reasoning*. It is often confused with **imply**, which means to *suggest* or *insinuate*.

> We can infer from the large population density that there is a large demand for services.

> The large population density implies that there is a high demand for services.

inflammable, flammable, non-flammable. Despite its **in-** prefix, **inflammable** is not the opposite of **flammable**: both words describe things that are *easily set on fire*. The opposite of **flammable** is **non-flammable**. To prevent any possibility of confusion, it's best to avoid **inflammable** altogether.

irregardless. Irregardless is non-standard English; use *regardless*.

its, it's. Its is a form of possessive pronoun; **it's** is a contraction of *it is*. Many people mistakenly put an apostrophe in **its** in order to show possession.

> ✗ The cub wanted it's mother.

> ✓ The cub wanted its mother.

> ✓ It's time to leave.

less, fewer. Less is used when units are *not* discrete and *not* absolute (as in "less information"). **Fewer** is used when the units *are* discrete and absolute (as in "fewer details").

lie, lay. To **lie** means to *assume a horizontal position*; to **lay** means to *put down*. The changes of tense often cause confusion:

Present	Past	Past participle
lie	lay	lain
lay	laid	laid

like, as. **Like** is a preposition, but it is often wrongly used as a conjunction. To join two independent clauses, use the conjunction **as**:

   ✗   I want to progress <u>like</u> you have this year.

   ✓   I want to progress <u>as</u> you have this year.

   ✓   Prof. Dimitriou is <u>like</u> my old school principal.

might of. See **could of.**

minimize. See **diminish.**

mitigate, militate. To **mitigate** means to *reduce the severity* of something; to **militate** against something means to *oppose* it.

myself, me. **Myself** is an intensifier of, not a substitute for, *I* or *me*:

   ✗   He gave it to John and <u>myself</u>.

   ✓   He gave it to John and <u>me</u>.

   ✗   Jane and <u>myself</u> are invited.

   ✓   Jane and <u>I</u> are invited.

   ✓   <u>Myself</u>, I would prefer a swivel chair.

nor, or. Use **nor** with **neither** and **or** by itself or with **either**:

      He is <u>neither</u> overworked <u>nor</u> underfed.

      The plant is <u>either</u> diseased <u>or</u> dried out.

off of. Remove the unnecessary **of**:

   ✗   The fence kept the children <u>off of</u> the premises.

   ✓   The fence kept the children <u>off</u> the premises.

phenomenon. A singular noun: the plural is **phenomena**.

plaintiff, plaintive. A **plaintiff** is a person who brings a case against someone else in court; **plaintive** is an adjective meaning *sorrowful*.

populace, populous. Populace is a noun meaning the *people* of a place; **populous** is an adjective meaning *thickly inhabited*:

> The <u>populace</u> of Hilltop village is not well educated.

> With so many people in such a small area, Hilltop village is a <u>populous</u> place.

practice, practise. Practice can be a noun or an adjective; **practise** is always a verb. Note, however, that the verb may also be spelled practice:

> The soccer players need <u>practice</u>. (noun)

> That was a <u>practice</u> game. (adjective)

> The players need to <u>practise</u> (or <u>practice</u>) their skills. (verb)

precede, proceed. To **precede** is to *go before* (earlier) or *in front of* others; to **proceed** is to *go on* or *ahead*:

> The faculty will <u>precede</u> the students into the hall.

> The medal winners will <u>proceed</u> to the front of the hall.

prescribe, proscribe. These words are sometimes confused, although they have quite different meanings. **Prescribe** means to *advise the use of* or *impose authoritatively*. **Proscribe** means to *reject, denounce,* or *ban*:

> The professor <u>prescribed</u> the conditions under which the equipment could be used.

> The student government <u>proscribed</u> the publication of unsigned editorials in the newspaper.

principle, principal. Principle is a noun meaning a *general truth* or *law*; **principal** can be used as either a noun or an adjective, meaning *chief*.

rational, rationale. Rational is an adjective meaning *logical* or *able to reason*. **Rationale** is a noun meaning *explanation*:

> That was not a <u>rational</u> decision.

> The president sent around a memo explaining the <u>rationale</u> for her decision.

real, really. Real, an adjective, means *true* or *genuine;* **really,** an adverb, means *actually*, *truly*, *very*, or *extremely.*

> The nugget was <u>real</u> gold.

> The nugget was <u>really</u> valuable.

seasonable, seasonal. Seasonable means *usual* or *suitable for the season;* **seasonal** means *of, depending on,* or *varying with the season:*

> It's quite cool today, but we can expect the return of <u>seasonable</u> temperatures later this week.

> You must consider <u>seasonal</u> temperature changes when you pack for such a long trip.

should of. See **could of.**

their, there. Their is the possessive form of the third-person plural pronoun. **There** is usually an adverb, meaning *at that place* or *at that point:*

> They parked <u>their</u> bikes <u>there</u>.

> <u>There</u> is no point in arguing with you.

tortuous, torturous. The adjective **tortuous** means *full of twists and turns* or *circuitous.* **Torturous**, derived from *torture*, means *involving torture* or *excruciating:*

> To avoid heavy traffic, they took a <u>tortuous</u> route home.

> The concert was a <u>torturous</u> experience for the audience.

translucent, transparent. A **translucent** substance permits light to pass through, but not enough for a person to see through it; a **transparent** substance permits light to pass unobstructed, so that objects can be seen clearly through it.

turbid, turgid. Turbid, with respect to a liquid or colour, means *muddy, not clear,* or (with respect to literary style) *confused.* **Turgid** means *swollen, inflated,* or *enlarged,* or (again with reference to literary style) *pompous* or *bombastic.*

unique. This word, which means *of which there is only one* or *unequalled*, is both overused and misused. Since there are no degrees of comparison—one thing cannot be "more unique" than another—expressions such as *very unique* or *quite unique* are incorrect.

while. To avoid misreading, use **while** only when you mean *at the same time that*. Do not use **while** as a substitute for *although*, *whereas*, or *but*:

    ✗   While she's getting fair marks, she'd like to do better.

    ✗   I headed for home, while she decided to stay.

    ✓   He fell asleep while he was reading.

-wise. Never use **-wise** as a suffix to form new words when you mean *with regard to*:

    ✗   Sales-wise, the company did better last year.

    ✓   The company's sales increased last year.

your, you're. **Your** is a pronominal adjective used to show possession; **you're** is a contraction of *you are*:

       You're likely to miss your train.

analysis of variance. A statistical test that compares the means of several samples to determine whether the difference between them could have occurred by chance.

applied research. Research intended to provide decision makers with practical, action-oriented recommendations to solve a problem.

baseline. A measure of conditions or behaviour before experimental manipulation is carried out.

basic research. Research intended to make and test theories about some aspect of real life.

between-subject variables. Experimental treatment conditions in which each group of subjects receives a different treatment. (Compare **within-subject variables**.)

confidence limits (and **confidence intervals**). A range of values expressing the likelihood that the mean estimated from a sample represents the true mean of a population. A "95 per cent confidence limit" means that there is only a 5 per cent chance that the true value is *not* included within the span of the error.

control (for). Examine the influence on the dependent variable of changes in one independent variable while holding constant (i.e., *controlling for*) other independent variables.

control group. Those subjects who are not exposed to the experimental manipulations. (Compare **experimental group**.)

deduce. Infer by reasoning from known facts.

dependent variable. What is measured in an experiment. This is what you expect to change as a result of an experimental manipulation. (Compare **independent variable**.)

experimental group. Those subjects who are exposed to the experimental manipulations. (Compare **control group**.)

F-ratio. The test statistic in an analysis of variance whose value will indicate whether the differences between means are statistically significant.

hypothesis. A statement of an expected relationship between two or more variables.

independent variable. What is manipulated in an experiment. This is the experimental treatment that is imposed on a subject to produce a change in behaviour. (Compare **dependent variable**.)

intervening variable. A characteristic or condition that explains the link between a cause and an effect; a variable through which the independent variable acts on the dependent variable.

logarithm. The exponent or power to which a base must be raised to yield a given number. So, for common, or base-10, logarithms, the value of the logarithm is the power to which 10 must be raised to give that number: e.g., $10^2 = 100$, therefore $\log_{10} 100 = 2$. A very wide range of numbers is sometimes converted to logarithmic values to compress the scale.

mean. The average of a set of scores, calculated as the total of all the scores divided by the number of scores in the sample.

model. A theoretical "picture" of the relations among causes and effects.

population. The entire collection of individuals from which a sample is drawn. In a typical psychology experiment in which first-year students are the participants, the population might be first-year university students, or all university students, or all individuals between 18 and 30 years of age. Which of these applies would depend on whether the individuals in the sample may be assumed to possess the relevant characteristics that may be generalized to the broader population.

qualitative data. Data that cannot be satisfactorily described by numbers and must be described in words.

quantitative data. Data that can be satisfactorily described by numbers.

regression line. A "line of best fit," representing the trend of a set of data. It is calculated using a regression equation and represents the line that passes closest to all the points in a data set. Depending upon the trend (i.e., linear or curvilinear) of the data, different regression equations may be used.

sample. The subset of individuals who are tested in an experiment and whose data may be generalized to a wider population.

significance (or **statistical significance**). The likelihood that an observed relationship has occurred by chance alone.

significance level. An estimate of the likelihood that the obtained result could have occurred by chance. A result that is significant at the .05 level indicates that the probability of a chance result is 5 out of 100.

standard deviation. A measure of the variability of scores in a sample. It tells you how closely a set of scores cluster around the mean of a set of data and is calculated as the square root of the average squared deviation of each score from the mean.

standard error (or **standard error of the mean**). An estimate of the variability of scores in a population, based on data from a single sample. Mathematically, it is the standard deviation of the sampling distribution of means.

theory. A set of interconnected statements or propositions that attempts to explain a causal relationship.

within-subject variables. Experimental treatment conditions in which a single group of subjects receives each different treatment. (Compare **between-subject variables**.)

GLOSSARY II: GRAMMAR

abstract. A summary accompanying a formal scientific report or paper, briefly outlining the contents.

abstract language. Language that deals with theoretical, intangible concepts or details: e.g., *justice*; *goodness*; *truth*. (Compare **concrete language**.)

acronym. A pronounceable word made up of the first letters of the words in a phrase or name: e.g., *NATO* (from *North Atlantic Treaty Organization*). A group of initial letters that are pronounced separately is an **initialism**: e.g., *CBC*; *NHL*.

active voice. See **voice**.

adjectival phrase (or **adjectival clause**). A group of words modifying a noun or pronoun: e.g., *the dog that belongs to my brother*.

adjective. A word that modifies or describes a noun or pronoun: e.g., *red*; *beautiful*; *solemn*.

adverb. A word that modifies or qualifies a verb, adjective, or adverb, often answering a question such as *how? why? when?* or *where?*: e.g., *slowly*; *fortunately*; *early*; *abroad*. (See also **conjunctive adverb**.)

adverbial phrase (or **adverbial clause**). A group of words modifying a verb, adjective, or adverb: e.g., *The dog ran with great speed*.

agreement. Consistency in tense, number, or person between related parts of a sentence: e.g., between subject and verb, or noun and related pronoun.

ambiguity. Vague or equivocal language; meaning that can be taken two ways.

antecedent (or **referent**). The noun for which a following pronoun stands: e.g., *cats* in *Cats are happiest when they are sleeping*.

appositive. A word or phrase that identifies a preceding noun or pronoun: e.g., *Mrs Jones, my aunt, is sick*. The second phrase is said to be **in apposition to** the first.

article. See **definite article**, **indefinite article**.

assertion. A positive statement or claim: e.g., *The data are inconclusive.*

auxiliary verb. A verb used to form the tenses, moods, and voices of other verbs: e.g., "am" in *I am swimming.* The main auxiliary verbs in English are *be, do, have, can, could, may, might, must, shall, should,* and *will.*

bibliography. 1. A list of works used or referred to in writing an essay or report.

2. A reference book listing works available on a particular subject.

case. Any of the inflected forms of a pronoun (see **inflection**).

Subjective case: *I, we, he, she, it, they.*

Objective case: *me, us, him, her, it, them.*

Possessive case: *my, our, his, her, its, their.*

circumlocution. A roundabout or circuitous expression, often used in a deliberate attempt to be vague or evasive: e.g., *in a family way* for "pregnant"; *at this point in time* for "now."

clause. A group of words containing a subject and predicate. An **independent clause** can stand by itself as a complete sentence: e.g., *I bought a hamburger.* A **subordinate** (or **dependent**) **clause** cannot stand by itself but must be connected to another clause: e.g., *Because I was hungry, I bought a hamburger.*

cliché. A phrase or idea that has lost its impact through overuse and betrays a lack of original thought: e.g., *slept like a log; gave 110 per cent.*

collective noun. A noun that is singular in form but refers to a group: e.g., *family; team; jury.* It may take either a singular or plural verb, depending on whether it refers to individual members or to the group as a whole.

comma splice. See **run-on sentence**.

complement. A completing word or phrase that usually follows a linking verb to form a **subjective complement**: e.g., (1) *He is my father*; (2) *That cigar smells terrible.* If the complement is an adjective it is sometimes called a **predicate adjective**. An **objective complement** completes the direct object rather than the subject: e.g., *We found him honest and trustworthy.*

complex sentence. A sentence containing a dependent clause as well as an independent one: e.g., *I bought the ring, although it was expensive.*

compound sentence. A sentence containing two or more independent clauses: e.g., *I saw the accident and I reported it.* A sentence is called **compound-complex** if it contains a dependent clause as well as two independent ones: e.g., *When the fog lifted, I saw the accident and I reported it.*

conclusion. The part of an essay in which the findings are pulled together or the implications revealed so that the reader has a sense of closure or completion.

concrete language. Specific language that deals with particular details: e.g., *red corduroy dress; three long-stemmed roses.* (Compare **abstract language**.)

conjunction. An uninflected word used to link words, phrases, or clauses. A **coordinating conjunction** (e.g., *and, or, but, for, yet*) links two equal parts of a sentence. A **subordinating conjunction**, placed at the beginning of a subordinate clause, shows the logical dependence of that clause on another: e.g., (1) *Although I am poor, I am happy*; (2) *While others slept, he studied.* **Correlative conjunctions** are pairs of coordinating conjunctions (see **correlatives**).

conjunctive adverb. A type of adverb that shows the logical relation between the phrase or clause that it modifies and a preceding one: e.g., (1) *I sent the letter; it never arrived, however.* (2) *The battery died; therefore the car wouldn't start.*

connotation. The range of ideas or meanings suggested by a certain word in addition to its literal meaning. Apparent synonyms, such as *poor* and *under-privileged*, may have different connotations. (Compare **denotation**.)

context. The text surrounding a particular passage that helps to establish its meaning.

contraction. A word formed by combining and shortening two words: e.g., *isn't* from "is not"; *we're* from "we are."

coordinate construction. A grammatical construction that uses correlatives.

coordinating conjunction. Each of a pair of correlatives.

copula verb. See **linking verb**.

correlatives (or **coordinates**). Pairs of coordinating conjunctions: e.g., *either/or; neither/nor; not only/but (also).*

dangling modifier. A modifying word or phrase (often including a participle) that is not grammatically connected to any part of the sentence: e.g., *Walking to school, the street was slippery.*

definite article. The word *the*, which precedes a noun and implies that it has already been mentioned or is common knowledge. (Compare **indefinite article**.)

demonstrative pronoun. A pronoun that points out something: e.g., (1) *This is his reason*; (2) *That looks like my lost earring*. When used to modify a noun or pronoun, a demonstrative pronoun becomes a kind of **pronominal adjective**: e.g., *this hat, those people*.

denotation. The literal or dictionary meaning of a word. (Compare **connotation**.)

dependent clause. See **clause**.

diction. The choice of words with regard to their tone, degree of formality, or register. Formal diction is the language of orations and serious essays. The informal diction of everyday speech or conversational writing can, at its extreme, become slang.

direct object. See **object**.

discourse. Talk, either oral or written. **Direct discourse** (or **direct speech**) gives the actual words spoken or written: e.g., *Donne said, "No man is an island."* In writing, direct discourse is put in quotation marks. **Indirect discourse** (or **indirect speech**) gives the meaning of the speech rather than the actual words. In writing, indirect discourse is not put in quotation marks: e.g., *He said that no one exists in an island of isolation.*

ellipsis marks. Three spaced periods indicating an omission from a quoted passage. At the end of a sentence use four periods.

essay. A literary composition on any subject. Some essays are descriptive or narrative, but in an academic setting most are expository (explanatory) or argumentative.

euphemism. A word or phrase used to avoid some other word or phrase that might be considered offensive or too harsh: e.g., *pass away* for *die*.

expletive. 1. A word or phrase used to fill out a sentence without adding to the sense.

2. A swear word.

exploratory writing. The informal writing done to help generate ideas before formal planning begins.

fused sentence. See **run-on sentence**.

general language. Language that lacks specific details; abstract language.

gerund. A verbal (part-verb) that functions as a noun and is marked by an *-ing* ending: e.g., *Swimming can help you become fit.*

grammar. The study of the forms and relations of words and of the rules governing their use in speech and writing.

hypothesis. A supposition or trial proposition made as a starting point for further investigation.

hypothetical instance. A supposed occurrence, often indicated by a clause beginning with *if*.

indefinite article. The word *a* or *an*, which introduces a noun and suggests that it is non-specific. (Compare **definite article.**)

independent clause. See **clause**.

indirect discourse (or **indirect speech**). See **discourse**.

indirect object. See **object**.

infinitive. A type of verbal not connected to any subject: e.g., *to ask*. The **base infinitive** omits the *to*: e.g., *ask*.

inflection. The change in the form of a word to indicate number, person, case, tense, or degree.

initialism. See **acronym**.

integrate. Combine or blend together.

intensifier (or **qualifier**). A word that modifies and adds emphasis to another word or phrase: e.g., *very tired*; *quite happy*; *I myself*.

interjection. An abrupt remark or exclamation, usually accompanied by an exclamation mark: e.g., *Oh dear! Alas!*

interrogative sentence. A sentence that asks a question: e.g., *What is the time?*

intransitive verb. A verb that does not take a direct object: e.g., *fall*; *sleep*; *talk*. (Compare **transitive verb**.)

introduction. A section at the start of an essay that tells the reader what is going to be discussed and why.

italics. Slanting type used for emphasis or to indicate the title of a book or journal.

jargon. Technical terms used unnecessarily or in inappropriate places: e.g., *peer-group interaction* for *friendship*.

linking verb (or **copula verb**). The verb *to be* used to join subject to complement: e.g., *The apples were ripe.*

literal meaning. The primary, or denotative, meaning of a word.

logical indicator. A word or phrase—usually a conjunction or conjunctive adverb—that shows the logical relation between sentences or clauses: e.g., *since*; *furthermore*; *therefore*.

misplaced modifier. A word or group of words that can cause confusion because it is not placed next to the element it should modify: e.g., *I only ate the pie.* [Revised: *I ate only the pie.*]

modifier. A word or group of words that describes or limits another element in the sentence.

mood. 1. As a grammatical term, the form that shows a verb's function.

Indicative mood: *She is going.*

Imperative mood: *Go!*

Interrogative mood: *Is she going?*

Subjunctive mood: *It is important that she go.*

2. When applied to literature generally, the atmosphere or tone created by the author.

non-restrictive modifier (or **non-restrictive element**). See **restrictive modifier**.

noun. An inflected part of speech marking a person, place, thing, idea, action, or feeling, and usually serving as subject, object, or complement. A **common noun** is a general term: e.g., *dog*; *paper*; *automobile*. A **proper noun** is a specific name: e.g., *Martin*; *Sudbury*; *Skidoo*.

object. 1. A noun or pronoun that completes the action of a verb is called a **direct object**: e.g., *He passed the puck*. An **indirect object** is the person or thing receiving the direct object: e.g., *He passed Simon* (indirect object) *the puck* (direct object).

2. The noun or pronoun in a group of words beginning with a preposition: e.g., *at the house*; *about her*; *for me*.

objective complement. See **complement**.

objectivity. A position or stance taken without personal bias or prejudice. (Compare **subjectivity**.)

outline. With regard to an essay or report, a brief sketch of the main parts; a written plan.

paragraph. A unit of sentences arranged logically to explain or describe an idea, event, or object. The start of a paragraph is sometime marked by indentation of the first line.

parallel wording. Wording in which a series of items has a similar grammatical form: e.g., *At her marriage my grandmother promised to love, to honour, and to obey her husband.*

paraphrase. Restate in different words.

parentheses. Curved lines enclosing and setting off a passage; not to be confused with square brackets.

parenthetical element. A word or phrase inserted as an explanation or afterthought into a passage that is grammatically complete without it: e.g., *My musical career, if it can be called that, consisted of playing the triangle in kindergarten.*

participle. A verbal (part-verb) that functions as an adjective. Participles can be either **present**, usually marked by an *-ing* ending (e.g., *taking*), or **past** (e.g., *having taken*); they can also be **passive** (e.g., *being taken* or *having been taken*).

part of speech. Each of the major categories into which words are placed according to their grammatical function. Some grammarians include only function words (nouns, verbs, adjectives, and adverbs); others also include pronouns, prepositions, conjunctions, and interjections.

passive voice. See **voice**.

past participle. See **participle**.

periodic sentence. A sentence in which the normal order is inverted or in which an essential element is suspended until the very end: e.g., *Out of the house, past the grocery store, through the school yard, and down the railway tracks raced the frightened boy.*

person. In grammar, the three classes of personal pronouns referring to the person speaking (**first person**), the person spoken to (**second person**), and the person spoken about (**third person**). With verbs, only the third-person singular has a distinctive inflected form.

personal pronoun. See **pronoun**.

phrase. A unit of words lacking a subject-predicate combination, typically forming part of a clause. The most common kind is the **prepositional phrase—**

a unit consisting of a preposition and an object: e.g., *They are waiting <u>at the house</u>*.

plural. Indicating two or more in number. Nouns, pronouns, and verbs all have plural forms.

possessive case. See **case**.

prefix. An element placed in front of the root form of a word to make a new word: e.g., *pro-; in-; sub-; anti-.* (Compare **suffix**.)

preposition. A short word heading a unit of words containing an object, thus forming a **prepositional phrase**: e.g., <u>under</u> *the tree*, <u>before</u> *my time*.

pronoun. A word that stands in for a noun. A **personal pronoun** stands for the name of a person: *I; he; she; we; they;* etc.

punctuation. A conventional system of signs (e.g., comma, period, semicolon, etc.) used to indicate stops or divisions in a sentence and to make meaning clearer.

reference works. Sources consulted when preparing an essay or report.

referent. See **antecedent**.

reflexive verb. A verb that has an identical subject and object: e.g., *Isabel <u>taught herself</u> to skate*.

register. The degree of formality in word choice and sentence structure.

relative clause. A clause introduced by a relative pronoun: e.g., *The man <u>who came to dinner</u> is my uncle*.

relative pronoun. *Who, which, what, that,* or their compounds, used to introduce an adjective or noun clause: e.g., *the house <u>that</u> Jack built;* <u>whatever</u> *you say*.

restrictive modifier (or **restrictive element**). A phrase or clause that identifies or is essential to the meaning of a term: e.g., *The book <u>that my aunt gave me</u> is missing*. It should not be set off by commas. A **non-restrictive modifier** is not needed to identify the term and is usually set off by commas: e.g., *This book, <u>which my aunt gave me</u>, is one of my favourites*.

rhetorical question. A question asked and answered by a writer or speaker to draw attention to a point; no response is expected on the part of the audience: e.g., *<u>How significant are these findings?</u> In my opinion, they are extremely significant, for the following reasons. . . .*

run-on sentence. A sentence that goes on beyond the point where it should have stopped. The term covers both the **comma splice** (two sentences incorrectly joined by a comma) and the **fused sentence** (two sentences incorrectly joined without any punctuation).

sentence. A grammatical unit that includes both a subject and a predicate. The end of a sentence is marked by a period.

sentence fragment. A group of words lacking either a subject or a verb; an incomplete sentence.

simple sentence. A sentence made up of only one clause: e.g., *Joaquim climbed the tree.*

slang. Colloquial speech considered inappropriate for academic writing; it is often used in a special sense by a particular group: e.g., *stoked* for "excited"; *dis* for "show disrespect for."

split infinitive. A construction in which a word is placed between *to* and the base verb: e.g., *to completely finish.* Many still object to this kind of construction, but splitting infinitives is sometimes necessary when the alternatives are awkward or ambiguous.

squinting modifier. A kind of misplaced modifier that could be connected to elements on either side, making meaning ambiguous: e.g., *When he wrote the letter finally his boss thanked him.*

standard English. The English currently spoken or written by literate people and widely accepted as the correct and standard form.

subject. In grammar, the noun or noun equivalent with which the verb agrees and about which the rest of the clause is predicated: e.g., *They swim every day when the pool is open.*

subjective complement. See **complement**.

subjectivity. A stance that is based on personal feelings or opinions and is not impartial. (Compare **objectivity**.)

subjunctive. See **mood**.

subordinate clause. See **clause**.

subordinating conjunction. See **conjunction**.

subordination. Making one clause in a sentence dependent on another.

suffix. An element added to the end of a word to form a derivative: e.g., *prepare*, *preparation*; *sing*, *singing*. (Compare **prefix**.)

synonym. A word with the same dictionary meaning as another word: e.g., *begin* and *commence*.

syntax. Sentence construction; the grammatical arrangement of words and phrases.

tense. A set of inflected forms taken by a verb to indicate the time (i.e., past, present, future) of the action.

theme. A recurring or dominant idea.

thesis statement. A one-sentence assertion that gives the central argument of an essay.

topic sentence. The sentence in a paragraph that expresses the main or controlling idea.

transition word. A word that shows the logical relation between sentences or parts of a sentence and thus helps to signal the change from one idea to another: e.g., *therefore*; *also*; *however*.

transitive verb. A verb that takes an object: e.g., *hit*; *bring*; *cover*. (Compare **intransitive verb**.)

usage. The way in which a word or phrase is normally and correctly used; accepted practice.

verb. That part of a predicate expressing an action, state of being, or condition that tells what a subject is or does. Verbs are inflected to show tense (time). The principal parts of a verb are the three basic forms from which all tenses are made: the base infinitive, the past tense, and the past participle.

verbal. A word that is similar in form to a verb but does not function as one: a participle, a gerund, or an infinitive.

voice. The form of a verb that shows whether the subject acted (**active voice**) or was acted upon (**passive voice**): e.g., *He stole the money* (active). *The money was stolen by him* (passive). Only transitive verbs (verbs taking objects) can be passive.

Appendix

Weights, Measures, and Notation

The conversion factors are not exact unless so marked. They are given only to the accuracy likely to be needed in everyday calculations.

1. IMPERIAL AND AMERICAN, WITH METRIC EQUIVALENTS

Linear measure

1 inch	= 25.4 millimetres exactly
1 foot = 12 inches	= 0.3048 metre exactly
1 yard = 3 feet	= 0.9144 metre exactly
1 (statute) mile = 1,760 yards	= 1.609 kilometres
1 int. nautical mile	
= 1.150779 miles	= 1.852 km exactly

Square measure

1 square inch	= 6.45 sq. centimetres
1 square foot = 144 sq. in.	= 9.29 sq. decimetres
1 square yard = 9 sq. ft.	= 0.836 sq. metre
1 acre = 4,840 sq. yd.	= 0.405 hectare
1 square mile = 640 acres	= 259 hectares

Cubic measure

1 cubic inch	= 16.4 cu. centimetres
1 cubic foot = 1,728 cu. in.	= 0.0283 cu. metre
1 cubic yard = 27 cu. ft.	= 0.765 cu. metre

Capacity measure

Name	System	Equal to	Metric
fluid oz.	imperial	1/20 imp. pint	28.41 ml
	US (liquid)	1/16 US pint	29.57 ml
gill	imperial	1/4 pint	142.07 ml
	US (liquid)	1/4 pint	118.29 ml
pint	imperial	20 fl.oz.(imp.)	568.26 ml
	US (liquid)	16 fl.oz.(US)	473.18 ml
	US (dry)	1/2 quart	550.61 ml
quart	imperial	2 pints	1.1365 litres
	US (liquid)	2 pints	0.9464 litre
	US (dry)	2 pints	1.1012 litres
gallon	imperial	4 quarts	4.546 litres
	US (liquid)	4 quarts	3.785 litres
peck	imperial	2 gallons	9.092 litres
	US (dry)	8 quarts	8.810 litres
bushel	imperial	4 pecks	36.369 litres
	US (dry)	4 pecks	35.239 litres

Avoirdupois weight

1 grain	= 0.065 gram
1 dram	= 1.772 grams
1 ounce = 16 drams	= 28.35 grams
1 pound = 16 ounces	
= 7,000 grains	= 0.45359237
	kilogram exactly
1 stone = 14 pounds	= 6.35 kilograms
1 quarter = 2 stones	= 12.70 kilograms
1 hundredweight = 4 quarters	
= 112 lb.	= 50.80 kilograms
1 (long) ton = 20 cwt. = 2,240 lb.	= 1.016 tonnes
1 short ton = 2,000 pounds	= 0.907 tonne

2. METRIC, WITH IMPERIAL EQUIVALENTS

Linear measure

1 millimetre	= 0.039 inch
1 centimetre = 10 mm	= 0.394 inch
1 decimetre = 10 cm	= 3.94 inches
1 metre = 100 cm	= 1.094 yards
1 decametre = 10 m	= 10.94 yards
1 hectometre = 100 m	= 109.4 yards
1 kilometre = 1000 m	= 0.6214 mile

Square measure

1 square centimetre	= 0.155 sq. inch
1 square metre = 10 000 sq. cm	= 1.196 sq. yards
1 are = 100 sq. metres	= 119.6 sq. yards
1 hectare = 100 ares	= 2.471 acres
1 square kilometre = 100 ha	= 0.386 sq. mile

Cubic measure

1 cubic centimetre	= 0.061 cu. inch
1 cubic metre = one million cu. cm	= 1.308 cu. yards

Capacity measure

1 millilitre	= 0.002 pint (imperial)
1 centilitre = 10 ml	= 0.018 pint
1 decilitre = 100 ml	= 0.176 pint
1 litre = 1000 ml	= 1.76 pints
1 decalitre = 10 l	= 2.20 gallons (imperial)
1 hectolitre = 100 l	= 2.75 bushels (imperial)

Weight

1 milligram	= 0.015 grain
1 centigram = 10 mg	= 0.154 grain
1 decigram = 100 mg	= 1.543 grain
1 gram = 1000 mg	= 15.43 grain
1 decagram = 10 g	= 5.64 drams
1 hectogram = 100 g	= 3.527 ounces
1 kilogram = 1000 g	= 2.205 pounds
1 tonne (metric ton) = 1000 kg	= 0.984 (long) ton

3. SI UNITS

Base units

Physical quantity	Name	Abbr. or symbol
length	metre	m
mass	kilogram	kg
time	second	s
electric current	ampere	A
temperature	kelvin	K
amount of substance	mole	mol
luminous intensity	candela	cd

Supplementary units

Physical quantity	Name	Abbr. or symbol
plane angle	radian	rad
solid angle	steradian	sr

Derived units with special names

Physical quantity	Name	Abbr. or symbol
frequency	hertz	Hz
energy	joule	J
force	newton	N
power	watt	W
pressure	pascal	Pa
electric charge	coulomb	C
electromotive force	volt	V
electric resistance	ohm	Ω
electric conductance	siemens	S
electric capacitance	farad	F
magnetic flux	weber	Wb
inductance	henry	H
magnetic flux density	tesla	T
luminous flux	lumen	lm
illumination	lux	lx

4. TEMPERATURE

Celsius (or Centigrade): Water boils (under standard conditions) at 100° and freezes at 0°

Fahrenheit: Water boils at 212° and freezes at 32°

Kelvin: Water boils at 373.15 kelvins and freezes at 273.15 kelvins.

Celsius	Fahrenheit
-17.8°	0°
-10°	14°
0°	32°
10°	50°
20°	68°
30°	86°
40°	104°
50°	122°
60°	140°
70°	158°
80°	176°
90°	194°
100°	212°

To convert Celsius into Fahrenheit: multiply by 9, divide by 5, and add 32.

To convert Fahrenheit to Celsius: subtract 32, multiply by 5, and divide by 9.

5. METRIC PREFIXES

	Abbr. or symbol	Factor
deca-	da	10
hecto-	h	10^2
kilo-	k	10^3
mega-	M	10^6
giga-	G	10^9
tera-	T	10^{12}
peta-	P	10^{15}
exa-	E	10^{18}
deci-	d	10^{-1}
centi-	c	10^{-2}
milli-	m	10^{-3}
micro-	μ	10^{-6}
nano-	n	10^{-9}
pico-	p	10^{-12}
femto-	f	10^{-15}
atto-	a	10^{-18}

These prefixes may be applied to any units of the metric system: hectogram (abbr. hg) = 100 grams; kilowatt (abbr. kW) = 1000 watts; megahertz (MHz) = 1 million hertz; centimetre (cm) = $^1/_{100}$ metre; microvolt (μV) = one millionth of a volt; picofarad (pF) = 10^{-12} farad, and are sometimes applied to other units (megabit).

6. POWER NOTATION

This expresses concisely any power of ten (any number that is composed of factors of 10). 10^2 or ten squared = $10 \times 10 = 100$; 10^3 or ten cubed = $10 \times 10 \times 10 = 1,000$. Similarly, $10^4 = 10,000$ and $10^{10} = 1$ followed by ten zeros = 10,000,000,000. Proceeding in the opposite direction, dividing by ten and subtracting one from the index, we have $10^2 = 100$, $10^1 = 10$, $10^0 = 1$, $10^{-1} = ^1/_{10}$, $10^{-2} = ^1/_{100}$, and so on; $10^{-10} = 1/10,000,000,000$.

7. BINARY SYSTEM

Only two units (O and 1) are used, and the position of each unit indicates a power of two.

One to ten written in binary form:

	eights (2^3)	fours (2^2)	twos (2^1)	one
1				1
2			1	0
3			1	1
4	1	0	0	
5		1	0	1
6		1	1	0
7		1	1	1
8	1	0	0	0
9	1	0	0	1
10	1	0	1	0

i.e. ten is written as 1010 ($2^3 + 0 + 2^1 + 0$); one hundred is written as 1100100 ($2^6 + 2^5 + 0 + 0 + 2^2 + 0 + 0$).

Index